The Spatial Organisation of Urban Agriculture in the Global South

This book examines the role and position of urban agriculture in the spatial and functional structure of cities in the Global South.

In the face of dynamic urbanisation and negative consequences of climate change, one of the key challenges is not only how to provide food for the ever-growing urban population but also how to achieve urban sustainability and simultaneously reduce the negative impact of cities on the natural environment. These problems are particularly urgent in the metropolises of the Global South that are experiencing the greatest population growth while struggling with increasing social inequalities and the resulting uneven distribution of resources. Examining the role that urban agriculture can play in addressing these challenges, this book draws on three case study cities: Havana, Singapore and Kigali. The case studies, differing in socio-economic, spatial, political and environmental terms, exemplify diverse characteristics of urban agriculture in different geographical conditions. Drawing on fieldwork conducted in each city, this book also provides a unique perspective on the constraints in the development of urban agriculture and the use of its full potential for urban sustainability.

This book will appeal to students and scholars, as well as decision makers, interested in the issues of urban sustainability, food security, spatial development and alternative food systems.

Ada Górna is a senior assistant in the Faculty of Geography and Regional Studies at the University of Warsaw, Poland, where she lectures on urban studies and food systems. She holds a PhD in the field of social sciences in the discipline of socio-economic geography and spatial management from the University of Warsaw and has published in a range of journals, including *Land,* the *Journal of Baltic Studies, Planning Perspectives*, and *Miscellanea Geographica.*

Earthscan Food and Agriculture

Food Policy in the United Kingdom
An Introduction
Martin Caraher, Sinéad Furey and Rebecca Wells

Peasants, Capitalism, and the Work of Eric R. Wolf
Reviving Critical Agrarian Studies
Mark Tilzey, Fraser Sugden, and David Seddon

Peasants, Capitalism, and Imperialism in an Age of Politico-Ecological Crisis
Mark Tilzey and Fraser Sugden

Genome Editing and Global Food Security
Molecular Engineering Technologies for Sustainable Agriculture
Zeba Khan, Durre Shahwar, and Yasmin Heikal

University Engagement with Farming Communities in Africa
Community Action Research Platforms
Edited by Anthony Egeru, Megan Lindow, and Kay Muir Leresche

Globalisation and Livelihood Transformations in the Indonesian Seaweed Industry
Edited by Zannie Langford

Principles of Sustainable Aquaculture
Promoting Social, Economic and Environmental Resilience, 2nd edition
Stuart W. Bunting

The Spatial Organisation of Urban Agriculture in the Global South
Food Security and Sustainable Cities
Ada Górna

For more information about this series, please visit: www.routledge.com/books/series/ECEFA/

The Spatial Organisation of Urban Agriculture in the Global South

Food Security and Sustainable Cities

Ada Górna

Routledge
Taylor & Francis Group
LONDON AND NEW YORK

earthscan
from Routledge

First published 2024
by Routledge
4 Park Square, Milton Park, Abingdon, Oxon OX14 4RN

and by Routledge
605 Third Avenue, New York, NY 10158

Routledge is an imprint of the Taylor & Francis Group, an informa business

© 2024 Ada Górna

British Library Cataloguing-in-Publication Data
A catalogue record for this book is available from the British Library

ISBN: 978-1-032-54440-3 (hbk)
ISBN: 978-1-032-55271-2 (pbk)
ISBN: 978-1-003-42984-5 (ebk)

DOI: 10.4324/9781003429845

Typeset in Times New Roman
by codeMantra

for Krzysztof

Contents

Acknowledgements

I would like to thank a number of people without whom writing this book would not have been possible. First of all, I owe thanks to my supervisor, Professor Mirosława Czerny from the Faculty of Geography and Regional Studies of the University of Warsaw, who not only offered advice and extensive support while I was preparing this book, but has also been an example for me to follow since the beginning of my academic career – a model geographer and researcher.

I also wish to thank several people who have contributed to developing the concept of this book over the last years, as well as to the preceding field research and final compilation of the results with their advice, suggestions, comments and constructive criticism. My gratitude goes to the members of the Faculty who have supported me for many years. I want to thank three professors in particular – Professor Marcin Solarz, Professor Andrzej Kowalczyk and Professor Jerzy Makowski. My thanks also go to Professor Maria Skoczek as well as Professor Angelina Herrera Sorzano from the University of Havana, who helped me prepare the first and, at the same time, the most uncertain research trip to the capital city of Cuba.

Furthermore, I wish to offer my thanks to the reviewers of my doctoral thesis: Professor Daniela Szymańska from the Nicolaus Copernicus University in Toruń and Professor Andrzej Matczak from the University of Lodz, who suggested that I publish the thesis in book form. Their valuable notes and suggestions substantially enhanced its final version.

I am also grateful to the vice-rector for Research at the University of Warsaw, Professor Zygmunt Lalak, as well as the authorities of the Faculty of Geography and Regional Studies, in particular the Dean of the Faculty, Professor Maciej Jędrusik and Deputy Dean Sylwia Dudek-Mańkowska, PhD, for awarding University funds in order to translate the book from Polish into English. My thanks for coordinating the translation go to Zuzanna Grzegorczyk and Małgorzata Klein from Pracownia Językowa. I am very grateful to Ewa Mierzwiak, Joanna Papis and Iwona Orzechowska for their technical support in obtaining funding. I also wish to thank Hannah Ferguson and Katie Stokes from Routledge publishing house.

This work would not have been possible without my family, especially my parents, who have always cared for my education and encouraged me in all of my academic endeavours.

Last but not least, I would like to thank my husband – Krzysztof, to whom I dedicate this book. Your contribution to its preparation is hard to describe. Thank you for all our joint research trips, all evening discussions and "brainstorms" as well as for the day on one of the beaches on Santiago, Cabo Verde, where the idea to research urban agriculture was born.

1 Introduction

According to the 2019 report of the United Nations Department of Economic and Social Affairs (UN DESA), city residents make up more than 55% of the current global population. By 2050, this percentage may increase to as much as 68% (UN DESA 2019). In the face of continuous, dynamic urbanisation and negative consequences of climate change, one of the key challenges becomes the issue of not only how to ensure the food security of ever-growing city population but also how to simultaneously implement the principles of sustainable development and limit the negative impact of cities on the natural environment.

Already in the 18th century, in *An Essay on the Principle of Population*, T. Malthus forecast inevitable exhaustion of Earth's natural resources (Malthus 1798). Those fears were reiterated in 1972 in the report on "The Limits of Growth", commissioned by the Club of Rome (Meadows et al. 2013). Although neither of those catastrophic visions ultimately came to pass, the debate on unequal distribution of good, resulting from social and economic disproportions, is becoming increasingly frequent in academic discourse. Excessive burden on ecosystems, increasing demand for energy and adverse climate change, which are largely attributable to the world's huge metropolises, demonstrate the need to revise the current approach to city development and to start building sustainable urban systems. According to the estimates of the United Nations Human Settlements Programme (UN-Habitat), cities are responsible for 75% of greenhouse gas emissions even though they only cover 2% of the world's surface. The total emissions take into account both the gases produced in the cities themselves (e.g. transport, heating and air conditioning of buildings, industry) and those produced outside cities in order to meet the cities' needs (e.g. power generation, as well as production of food and other goods) (UN-Habitat 2011). In view of the aforementioned challenges, there have been proposals for the development of compact cities as well as short-distance cities that make up a functional whole. In such a city, it would be possible to shorten the distance between the place of production and place of consumption, while the residents would be able to satisfy their basic needs without having to travel by car. What is more, the new green cities should make efficient use of natural resources, as well as increase their energy and food self-sufficiency (Fücks 2013). Such development would limit negative impact of large metropolises on the natural environment while simultaneously improving the well-being of the residents themselves.

DOI: 10.4324/9781003429845-1

Currently, concepts such as smart city, green city, resilient city or soft city are gaining popularity. They all highlight – albeit in different ways – the need for thorough changes in the dominant paradigm of development of urban areas. Earth has become a planet of cities and it is cities that bear the great responsibility resulting from local processes taking place in them that have global, frequently catastrophic effects. The awareness of this fact as well as the effort to gradually transform urban systems are ingrained in the concept of intelligent and sustainable city development (Szymańska & Korolko 2015). While many aspect of city life require those changes, it is the issue of food production, distribution and consumption that is being more and more frequently tackled by scientists, planners and urban decision-makers.

The subject of this book is urban agriculture, which – when properly managed – presents a chance for sustainable development of cities. The financial crisis of 2007–2009,[1] the COVID-19 pandemic announced by the World Health Organisation in 2020 and even the war in Ukraine, the effects of which included a serious increase in food prices, only aggravated the issue of limited access to food products, especially among the poorest city residents. This problem is particularly serious in metropolises of the Global South, which are the focus of the considerations contained in this book. It is those cities that are experiencing the biggest increase in population while simultaneously struggling with pollution, growing social inequality and the resulting uneven distribution of resources. Local food production can serve as a remedy to the aforementioned problems. It presents a chance for improved food security of the growing urban population, increase in their self-sufficiency as well as enhancement of urban ecosystems. The book tackles the role played by urban agriculture in the spatial and functional structure of three cities of the Global South – Havana, Singapore and Kigali.

1.1   Outline of the research subject matter

The subject matter of the research described in the book is urban agriculture, which has been growing in significance since the 1990s. It is currently an important topic in research on sustainable urban development. The very term "urban agriculture" may seem like an oxymoron. After all, agriculture is commonly considered an activity strictly associated with rural areas, while the urbanisation process in economic terms is measured precisely with the share of population employed outside this sector of the economy. Moreover, crop growing and animal breeding in cities are perceived as archaic, temporary and marginal activities, which at best can be symptoms of a new fad that play additional recreational roles or improve the aesthetic of the urban landscape. However, in reality, urban agriculture is a prominent economic activity, which plays a key role in the lives of hundreds of millions of people around the world (Smit et al. 2001). It is estimated that in the mid-1990s, the number of city residents engaged in agriculture exceeded 200 million, while 800 million people were dependent on supplies of food produced by those workers (Zezza & Tasciotti 2010). According to various sources, the share of urban population engaged in agriculture at the time amounted to 40%–70% in Africa, 60% in Asia and 50% in Latin America (Bryld 2003; Zezza & Tasciotti 2010). Although those are merely estimates, they demonstrate the important role played by urban agriculture on a global scale, in particular in cities of the Global South. It is currently a rapidly developing sector, with growing significance for the food security of urban residents (in particular in the least developed cities) and a far-reaching economic, social and environmental impact.

Due to its comprehensive and complex nature, as well as a multitude of functions fulfilled in the city system, urban agriculture is the subject matter of research by representatives of many academic disciplines – agronomics, sociology, economics, urban studies, landscape architecture, spatial development as well as geography. The considerations contained in this book form a part of the contemporary trend of urban studies, research on sustainable development of cities as well as food security of their residents. Moreover, due to the comprehensive approach to urban agriculture and its place in the spatial and functional structure of the analysed cities, this study also contributes to research on urban food systems (Ingram 2011). This book combines three sub-disciplines of geography – socioeconomic geography, urban geography and agricultural geography. Apart from that, it contains spatial analyses that utilise the methodological apparatus of remote sensing and considerations that align with the research directions of political geography.

This book analyses four dimensions of urban agriculture: spatial, subjective, objective and functional. They were used here since they allow for a structured and exhaustive analysis of this comprehensive subject matter of research. The spatial dimension pertains to the location of urban agriculture within the analysed cities, the characteristics of its distribution and its spatial relations with other elements of the spatial and functional structure. Although the spatial aspect is particularly important, especially from the perspective of geographical research, it has been rarely tackled to date in available international literature. There also has been no comprehensive analysis of the spatial distribution of urban agriculture within the three selected cities. This book, which emphasises the spatial dimension of urban agriculture, seeks to fill the said gap in empirical research. Spatial analyses contained herein and the resulting cartographic studies present the distribution of agriculture within the space of the analysed cities, as well as the spatial organisation of individual urban gardens and farms and entire intra-urban agricultural areas. The three terms – urban gardens, urban farms and agricultural areas – are used in this book for a good reason. In each of the analysed cities, agriculture takes different forms that require different labels. In Havana's case, they are urban gardens (terminology will be discussed more broadly in Chapter 3), in the case of Singapore – urban farms (apart from a single instance of a community garden), and in the case of Kigali – agricultural areas (alternatively, crop fields or household gardens).

Other dimensions of urban agriculture analysed in the book are subjective and objective dimensions. The former takes into account the actors involved in urban agriculture, including producers, vendors, intermediaries, suppliers as well as consumers. The latter pertains to the products of urban agriculture and methods used for their production. The final dimension – functional one – covers the benefits provided by urban agriculture, from economic through social to environmental ones, as well as its role in the spatial and functional structure of the analysed cities.

The aforementioned dimensions were taken from a study by W. Sroka (2014). The author also proposes a fifth dimension – dichotomous one, that is, juxtaposing urban and rural agriculture. Although the differences between the two will be pointed out in this book, its empirical part focuses primarily on urban agriculture, specifically intra-urban agriculture, so including the dichotomous dimension in the framework of the conducted research is not required.

Apart from the dimensions discussed above, the book also presents the institutional and legal framework of urban agriculture that plays a key role in shaping its internal features and the features of its distribution. This framework comprises the applicable legal regulations, in particular those pertaining to the land ownership system, as well as planning and strategic documents presenting the directions of the spatial development of the analysed cities. They are the result of the policy pursued by central and municipal authorities, whose competencies include assignment of the ownership rights, determination of terms of land lease and development of planning documents designating the areas where agricultural activity (i.e. growing plants or breeding animals) is possible or desirable. Legal regulations also have a direct impact on internal features of urban agriculture, such as selection of production methods and techniques. Using appropriate regulations, the authorities can, for example, prohibit the use of artificial fertilisers or pesticides and apply incentives addressed to farm and garden owners as well as individual farmers to convince them to use the specified production organisation models. Using the institutional and legal framework, the authorities therefore determine the features and functions of urban agriculture; however, they can only do it in a given environmental, socioeconomic and political context. This is why distribution of urban gardens and farms, even though it can be influenced by the authorities, primarily depends on access to natural resources such as land and water, as well as on terrain, soil quality and climate conditions. In order to paint a fuller picture of urban agriculture, this book also presents the broader context in which it operates in each analysed metropolis.

This publication analyses the role played by urban agriculture in the three selected cities of the Global South. Therefore, the book adopts a functional and structural approach since the analysis covers the role of the element (urban agriculture) in the functioning of the whole (city), as well as relationships between urban agriculture and other elements of the spatial and functional structure of selected cities. The presented study is primarily empirical and cognitive in nature. This book relies on three case studies and is meant to improve the state of knowledge on the features of urban agriculture and the role it plays in the spatial and functional structure of cities of the Global South. Nevertheless, the conducted research also has a methodological dimension. Its results allow for designation of methods that are suitable for analysis of urban agriculture functioning in different socioeconomic, political and environmental conditions.

1.2   Spatial coverage and temporal scope

The considerations included herein concern cities of the Global South. The subject matter of detailed research includes three selected cases – Havana (the capital city of Cuba), Singapore and Kigali (the capital city of Rwanda). There are many methods of dividing the world into two regions – Global North and Global South. However, the most frequently quoted division is the one along the so-called "Brandt Line", proposed in 1980 by the Independent Commission for International Developmental Issues, led by the former Federal Chancellor of the Federal Republic of Germany and Nobel Peace Prize laureate Willy Brandt. Due to its common application, this line has become one of the best-known and simultaneously influential spatial depictions of the global developmental division. However, it is also a source of numerous controversies (Solarz 2012, 2019). First of all, they arise from the erroneous assumption that the countries assigned to those two separate groups form coherent, homogenous wholes. Another important issue is the multitude of divisions of the world based on the level of socioeconomic development (Solarz 2012, 2014, 2019).

The Global South includes more than 100 countries, inhabited by more than three quarters of the world's population. Those countries are located on different continents and in different climate zones; they vary considerably in terms of access to resources, affluence of the residents, political regime, culture as well as international policy directions. Due to the differences between the countries of the region, the criteria of dichotomous division of the world are dubious. First and foremost, the assumption that one group (the Global North) only includes highly developed countries, while the other one (the Global South) – exclusively the poorly developed ones, is debatable.

On the basis of data contained in the Human Development Report 2021–2022, published in 2022 by the United Nations Development Programme (UNDP), countries which are commonly assigned to the latter group belong to all categories of countries distinguished on the basis of the Human Development Index (HDI) – with low, medium, high and very high level of development (UNDP 2022). Although countries classified as the Global South are indeed the most numerous group among countries with low and medium levels of social development, they also make up 81.3% and 33.3%, respectively, among countries with high and very high levels of development. Examples of countries from the Global South which are characterised by very high level of social development include Singapore, Israel, South Korea, United Arab Emirates, Saudi Arabia, Bahrain, Chile, Qatar, Argentina, Brunei and Uruguay.

Apart from the socioeconomic criterion, measured using the aforementioned HDI, another determinant of the division into the Global North and Global South is the scope of political rights and civic freedoms (Solarz 2012, 2019). It is measured using, among others, the "Freedom in the World" index, according to which the non-governmental organisation Freedom House classifies countries of the world in three categories: free country, partly free country and not free country. According to the 2023 report (Freedom House 2023), countries of the Global South which are considered free include, for instance, Brazil, Chile, Uruguay, Argentina, Panama, Botswana, Namibia, Ghana and South Korea. In turn, partly free countries are, among others, Peru, Bolivia, Haiti, Senegal, Nigeria, Tunisia, Tanzania, Kenya, India, Indonesia and Malaysia. The last group – countries which are completely not free – include Cuba, Venezuela, Zimbabwe, Uganda, Rwanda, Ethiopia, Thailand, Vietnam and Singapore. Therefore, countries of the Global South have been classified in all three categories, which means that the region is also highly diversified in terms of the scope of political rights and civic freedoms. The fact that Cuba, Rwanda and Singapore are considered not free indicates that all three cases should be included in the region of Global South. The level of socioeconomic development is, after all, not the only criterion that should be taken into account in the division into countries of the Global North and Global South. This is particularly important in the case of Singapore, as its presence in the latter group may give rise to doubts due it its very high HDI.

The division into Global North and Global South, which has replaced the previous differentiation into First, Second and Third World countries as well as developed and developing countries is frequently criticised; however, it should be considered partially useful. It helps us understand, explain and structure the surrounding world to a certain extent. For this reason, it is commonly used in academic, political and media discourse (Solarz 2012).

Despite socioeconomic and political differences between the countries and cities of the Global South, there is no doubt that they also share a number of features. On a global scale, the metropolises of the region exhibit the fastest growth. The list of the largest cities in the world – the so-called megacities, whose population has exceeded ten million people – primarily includes those classified as cities of the Global South. According to the latest report of the UN DESA, published in 2019, 28 out of 33 megacities are metropolises of the Global South. Those megacities are, in descending order: Tokyo, Delhi, Shanghai, São Paulo, Mexico City, Mumbai, Beijing, Dhaka, Karachi, Buenos Aires, Chongqing, Istanbul, Kolkata, Manila, Lagos, Lima, Tianjin, Kinshasa, Canton, Shenzhen, Lahore, Bangalore, Bogotá, Jakarta, Chennai, and Bangkok (UN DESA 2019). Another characteristic feature of settlement systems in many countries of the region is the primate city phenomenon, which is a consequence of population concentration in large cities. Primate city is a term used to describe a leader in terms of size, dominating in the given country's settlement network. The phenomenon means concentration of the population in a given city in relation to the total population of a given country or a given city's population accounting for an overwhelming percentage of total urban population. In the report of the United Nations Department of Economic and Social Affairs (UN DESA 2019), the term "primate city" refers to settlement units inhabited by at least 40% of the country's total urban population. According to the document, as many as 20 of 27 primate cities are classified as belonging to the Global South. These are Hong Kong, Singapore, Asunción, Kuwait, Panama, Brazzaville, Monrovia, Montevideo, Lomé, Phnom Penh, Tel Aviv, Nouakchott, Kigali, Cairo, Beirut, Ouagadougou, Kabul, Port-au-Prince, Santiago and Lima (UN DESA 2019). Therefore, they include two of the cities analysed in this book – Kigali and Singapore.

Due to the pressure of the global market, metropolises of the region are unable to develop a spatial structure that would meet the needs and aspirations of their residents. Problems and challenges typical of the majority of cities of the Global South struggling with continuous population growth primarily include social stratification and the related spatial segregation, as well as establishment of marginal districts. Another typical feature of those metropolises is the important role played by the informal sector (in both construction and commerce). Nevertheless, cities of the region are more and more frequently undergoing processes that are also typical of highly developed countries. They include, in particular, urban sprawl and spread of housing estates inhabited by urban elites into suburban areas, as well as gentrification in the central districts (Czerny 2012). Despite the aforementioned similarities between metropolises of the Global South, it is important not to treat them in an arbitrary manner. Traits of individual cities comprise a number of socioeconomic and political conditions as well as global factors, as demonstrated by the research contained in this book.

It should be emphasised that the Global South is not treated here as a homogenous region, consisting of countries with a similar (low) development and wealth level. On the contrary, one of the assumptions of this study is to present yet another domain in which the Global South is internally diverse. Demonstrating the comprehensive and complex nature of urban agriculture itself is yet another proof of comprehensive and complex nature of the entire region.

The main part of this book comprises three case studies of urban agriculture in Havana, Singapore and Kigali. Their choice merits an explanation, since it was not accidental and resulted from a number of both substantive and practical motives. Cuba, Singapore and Rwanda exhibit different levels of socioeconomic development. According to the HDI ranking from 2022, among the 191 countries included in the ranking, Singapore is on the 12th place – in a group of countries with a very high level of development, Cuba on the 83rd place – in the group of countries with a high level of development, while Rwanda is on the 165th position – in the group of countries with a low level of development (UNDP 2022). These three countries are also characterised by different political systems – Rwanda is a republic, Cuba – a socialist republic, while Singapore – a republic with a semi-authoritarian system. Nevertheless, according to the division into the Global North and Global South, in the majority of academic studies, all three countries are classified in the latter group (among others, by Boniface 2003; Boyd & Comenetz 2007; Solarz 2012, 2014; Nouschi 2016; Solarz 2019). The city selection was based, among others, on the geographical location. They are situated in three different regions of the Global South – Havana in Latin America, Singapore in Southeast Asia and Kigali in Sub-Saharan Africa. This fact enabled an analysis of urban agriculture in diametrically different conditions and presentation of a broad spectrum of its features and the factors shaping it. The three cities differ not only in terms of the level of development and the political regime but also features of the spatial and functional structure itself. Their choice has enabled, on the one hand, specification of differences in the features of urban agriculture operating in diverse socioeconomic and political conditions, while, on the other hand, of universal attributes of urban agriculture, present regardless of different local conditions. In order to ensure comparability of the conducted research, capital cities with a similar area were chosen (Havana – 728.3 km², Singapore – 734.3 km², Kigali – 730 km²). Due to the prestige associated with the role they play, capital cities are typically characterised by more dynamic urban processes, as well as more intense competition for space between various actors, including those engaged in urban agriculture. Capital cities are typically where important administrative buildings are located – ministries and embassies, as well as banks, hotels, universities or headquarters of international corporations. Choosing three capital cities allowed for an analysis of the role of urban agriculture amongst intensified competition for urban space.

The selection of those cities also resulted from certain individual conditions, separate for each of the analysed cases. Each analysed city is unique in its region in terms of urban agriculture, which was one of the basic criteria of their selection. The first case study is Havana, the capital of Cuba and one of the largest metropolises in the Caribbean. Dynamic development of urban agriculture in that city only began in the 1990s as a result of the economic crisis caused by the downfall of the Eastern Bloc, which led to a serious reduction of food product supplies to the island, mainly from the USSR. Due to the ubiquity of urban gardens within the city space as well as the role they play in supplying the residents with food, Havana is one of the most interesting subjects of studies on urban agriculture in all of Latin America. Its choice offers an opportunity to examine the history of an alternative food system, based on organic production methods and developing in an environment with limited resources.

The second case study is Singapore, one of the fastest developing metropolises in the world, considered a representative example of a smart city. The ICT solutions implemented there are meant to improve the well-being of the residents, ensure sustainable use of natural resources as well as improve the quality of the natural environment. However, Singapore is facing a major challenge due to its nearly complete dependency on food imports from abroad. Therefore, the authorities of this Asian city-state have begun to see urban agriculture as a chance for improving food self-sufficiency. Nevertheless, due to the limited spatial resources, growing building pressure and increasing land prices, agriculture in Singapore has to take forms that allow for highly efficient use of space and, at the same time, bring economic profits sufficient for it to remain on the market. Therefore, this city should be, on the one hand, considered a laboratory of modern agriculture based on advanced technologies (whose experience is used by other cities in the region, such as Hong Kong, Shanghai, Kuala Lumpur, Taipei, Tokyo or Macau); on the other hand, it is a city where more traditional urban agriculture is losing its significance. Contrary to Havana, the choice of Singapore allows for the role of local food production to be analysed amongst nearly unlimited economic resources and dynamic development typical of metropolises of the region.

The last selected city is Kigali, the capital of Rwanda, which is representative of other metropolises of Sub-Saharan Africa in terms of distribution of urban agriculture. Similar to Bissau, Brazzaville, Yaoundé or Kampala, agriculture in Kigali is also concentrated in vast bottoms of valleys and occupies a high share of space, reflecting the nutritional needs, especially of the poorest social groups (Górna & Górny 2020). However, it should be noted that despite the fact that the capital of Rwanda is still struggling with a number of issues typical of other metropolises of the region, such as poverty, economic disproportions and expansion of marginal districts, it is also an example of a city whose authorities, like those in Singapore, are striving to increase the attractiveness of investments and simultaneously promote so-called green initiatives. The Rwandan capital is a unique example, since the research carried out there allowed for the role of urban agriculture to be outlined in a city on the verge of dynamic socioeconomic development. It created a chance to "capture" its features and functions on the "eve" of the upcoming changes resulting from the growing building pressure as well as spatial policy of the authorities (included in planning and strategic documents, such as the Kigali Master Plan, published in 2020), aimed at "structuring" the urban tissue.

On the basis of the three selected instances, it was possible to describe the role of urban agriculture in metropolises that exhibit different rates of changes taking place in their spatial and functional structure. Havana should be considered a stagnant city, where spatial expansion and intensive increase in building density is not observed – with the exception of the *Miramar* district, where new luxurious hotels have been constructed since Cuba opened up to tourists. However, undeveloped or abandoned parcels still remain within the district's space, which demonstrates the slow pace of increasing building density. In Singapore, "gaps" in space have already been filled as a result of dynamic development. Currently, due to a serious space deficit, this Asian city-state is using even the most peripheral areas for construction. It is also building artificial islands. On the other hand, Kigali exemplifies a city with remaining undeveloped land within its space. However, due to numerous investments in infrastructure, this land will likely be used for construction in the near future.

Since this book is based on comprehensive field research, the accessibility of the cities was also an important factor in their selection. It was necessary to choose locations where the planned research could be carried out. All three cities were and still remain sufficiently safe so that the fieldwork was successfully carried out even in marginal districts and none of the assumed activities presented any risk.

Taking into account the fact that the detailed research described in the book covers primarily intra-urban farming, that is, agriculture located in a densely developed urban area, it was necessary to limit the research area for two of the selected cases – Havana and Kigali. The administrative borders of both cities include vast nearby rural areas, which were excluded from this study. In the case of Singapore, whose administrative borders nearly completely overlap with the island's borders (the study only excluded several uninhabited islands, which are undeveloped or fulfil exclusively industrial or recreational functions) and whose densely developed area is not surrounded by rural areas, it was not necessary to reduce the spatial coverage of research.

The detailed analysis contained in this book covers predominantly the situation observed during the field research conducted in May 2018 in Havana, at the turn of January and February 2019 in Singapore and in July 2019 in Kigali. This book is therefore a snapshot of urban agriculture at a given point in time. Regardless of the foregoing, the considerations contained herein cover a much broader time range than the dates of the field visits. The characterisation of the spatial and functional structure of the selected cities, preceding each description of urban agriculture in a given city, always covers the time of their founding and even the first settlements in the current location, sometimes expanding the temporal perspective by several centuries (Havana was founded in 1519, Singapore in 1819 and Kigali in 1907). Moreover, the following statistical analysis of distribution of urban agriculture was broadened using elements of dynamic analysis. In Havana's case, this analysis dates back to 2000 (the date of the oldest available satellite image) in Singapore – to 1974 (the date of founding of the oldest analysed farm), and in Kigali – to 2019 (the date of field research). The date ending the considerations included in the book was in all cases the year 2022, from which the latest available satellite images or information published on official websites of urban farms and gardens come.

1.3   Research objectives, questions and hypotheses

For the purposes of this book, one main objective and five detailed objectives were formulated. They are presented below along with auxiliary research questions. Moreover, threeresearch hypotheses were also developed – they are verified in the final chapter.

The main objective of this book is to determine the characteristics of distribution, internal features and functions of urban agriculture within the spatial and functional structure of selected cities of the Global South. It was assumed that the internal (endogenous) features were the structural and production as well as organisational and technical features of individual urban gardens, urban farms and agricultural areas, while their external (exogenous) features were the environmental, socioeconomic and political conditions in which a given garden, farm or agricultural area operated. The characterisation of the external features will serve as the basis for explaining the causes affecting the internal features of the agriculture itself as well as the functions it fulfils. Differentiating between endogenous and exogenous features is particularly important in order to avoid erroneous identification of causes and effects.

Due to the complexity of the research subject matter tackled, for all three analysed cases, the following **detailed objectives** were formulated, whose achievement will allow for answering the research questions written in italics:

1 Indicating the locations and features of distribution of urban agriculture

- Where are individual urban gardens/farms, agricultural areas located?
- What land do they occupy (next to houses, public, urban wasteland zones, flat terrain, inclines [slopes] and valleys)?
- Is urban agriculture dispersed or concentrated in particular parts of cities?

2 Determining the factors affecting the distribution of agriculture

- What are the reasons for urban agriculture being located in individual parts of the city (e.g. supply of land resources, low land prices, presence of wasteland zones, proximity of the market)?
- What are the causes of concentration/dispersion of urban agriculture?

3 Indicating the characteristics of urban agriculture, including

a Structural and production characteristics:

- What is the structure of plant production?
- Which plant species are grown?
- Do food, industrial or fodder crops dominate?
- What is the estimated volume of plant production?
- What is the structure of animal production?

b Organisational and technical characteristics:

- Which production methods techniques are used?
- Are artificial fertilisers used?
- Is compost produced?
- What methods of protection against pests are used: natural/artificial?
- How many people are employed and on what terms?
- Is it a private or state-owned enterprise?

4 Describing the paths of the products from the place of production to the place of distribution/consumption

- Are the products sold or intended to meet the producers' own needs?
- Where are the products sold?
- Where are they transported to?
- What route do the products take?
- What is the distance between the place of production and place of distribution/consumption of the products?
- Are the products intended as the producers' own supply?

5 Determining the functions fulfilled by urban agriculture

- Does agriculture fulfil exclusively nutritional functions?
- Is agriculture multifunctional and, apart from nutritional function, does it also play social, environmental, educational and tourist roles?

With regard to the foregoing detailed objectives and research questions, the following three **research hypotheses** were formulated:

H1. Factors affecting the location of urban agriculture are not universal, but in each analysed case, they are the resultant of local socioeconomic, political and environmental conditions.

H2. Urban agriculture in the analysed cities operates within a shortened supply chain, both in spatial terms, understood as the distance between the place of production and place of distribution/consumption, as well as subjective, understood as the number of actors along the products' route.

H3. In each analysed case, urban agriculture fulfils different functions, depending on the socioeconomic and political conditions.

1.4   Research methods applied

The research procedure in this book was divided into three main stages: analysis and preparation (I), field work (II) and comparison and summary (III). These stages were carried out with respect to all three selected case studies – Havana, Singapore and Kigali. They are characterised below, along with the methods used during the implementation of each stage.

- Stage I (analysis and preparation)

Stage I included analysis of international academic literature, planning documents and online sources, allowing for selection of the case studies and detailed characterisation of their spatial and functional structure, as well as analysis of generally available satellite and aerial images, which enabled preliminary location of urban agriculture within the city space. Moreover, stage I included substantive and organisational preparation for the field research. It consisted in outlining the routes along which agriculture was mapped in each city, preparing questionnaire forms for data collection in the field and preparing a list of issues to be touched upon during semi-structured interviews.

The options of using remote sensing for urban agriculture research are very limited. This is mainly due to the insufficient spatial resolution of publicly available satellite images as well as the attributes of urban agriculture itself. First of all, a typical feature of the structure of crops located within cities is high heterogeneity of species. Various plant species are frequently grown in a single garden, farm, crop field or even plant bed. Second, due to the fact that the objective of urban agriculture is to ensure continuous food supplies, plants grown next to each other can be at different stages of development. In both cases, plants within a single pixel have therefore different spectral characteristics, which rules out the application of automated remote sensing tools. Third, agriculture very frequently occupies small spaces, up to about a dozen square metres. Therefore, the publicly available (and free) satellite images obtained by satellites like Sentinel and Landsat are characterised by spatial resolution (measured by pixel length, i.e. the smallest distinguishable digital unit in a given image) insufficient to identify urban agriculture and specify its characteristics with the assistance of automated methods. For example, spatial resolution of Sentinel-2 images is up to 10 m, which means that the smallest crop field recognisable in them (assuming that it has uniform land coverage) must occupy at least 100 m^2. For that reason, use of automated remote sensing methods was omitted in this book. Instead, the chosen method was manual analysis of the images available in Google Earth Pro, an application which has been successfully used in similar research carried out by other authors in Hanoi (Forster, Buehler & Kellenberger 2009), Chicago (Taylor & Lovell 2014), Rome (Pulighe & Lupia 2016) or in Nakhon Ratchasima in Thailand (Jantakat et al. 2019). The resolution of the orthophotomap (developed on the basis of satellite and aerial images) is up to 30 cm, thanks to which its use in analysis of small fields with diverse spectral characteristics has proven effective. On the basis of the images available in the programme, densely developed research areas were selected (for Havana and Kigali). Within those areas, manual and visual interpretation was carried out, which allowed for identification of the polygons occupied by urban agriculture. Account was taken of direct identifying features, such as shape, size, structure and colour, as well as indirect features, such as the shade cast by facilities within the analysed polygons, their location, as well as associations with other landscape features. The margin of error was set at 5%, which is typical of research in social sciences. During the analysis, the polygon tool was also used to mark the identified gardens, urban farms and crop fields as well as to measure the area taken up by them. Moreover, every site was also assigned an ID number. The data of the sites from Google Earth Pro (containing the coordinates of the polygons identified as urban agriculture and designated routes) in the .kmz format were input in the MAPS. ME application, which was subsequently used to collect data in the field. MAPS. ME is a mobile application that shares maps from the OpenStreetMap portal and features a GPS tool, allowing for one's location to be determined in the field. Its major advantage is the fact that it can be used offline. Internet access in the visited cities (especially in Havana and Kigali) was limited, so the application became an important work tool, allowing for the identified sites to be easily found in the field.

• Stage II (fieldwork)

Stage II covered field research carried out in the three selected cities (in Havana in May 2018, in Singapore at the turn of January and February 2019 and in Kigali in July 2019). It involved mapping of urban agriculture within the previously designated research areas. During the field research, semi-structured interviews with actors involved in urban agriculture were conducted, including interviews with farm/garden owners, crop fieldworkers and vendors at local markets, as well as decision-makers in urban agriculture management. A particularly important element of the fieldwork was standardised field observation of the agricultural areas or urban farms and gardens as well as their immediate vicinity and the accompanying collection of photographic documentation. Stage II also included substantive consultations with representatives of local academic units dealing with urban agriculture. Thus, academic contact was established with representatives of the university in Havana (*Universidad de la Habana*) and the Kigali Independent University. The field research allowed for verification of the results and effectiveness of research methods applied at stage I.

The following is a detailed description of the research methods applied at stage II:

- Mapping of urban agriculture – it was carried out according to the routes designated during stage I. The sites previously identified in Google Earth Pro and entered in the mobile application MAPS.ME were located in the field using a GPS tool (available offline in the application).
- Standardised field observations – they were carried out according to the previously prepared questionnaire form (paper or electronic one). During those observations, particular attention was paid to the structure of production (plant or animal), production methods and techniques used within the analysed facility, as well as features of the surrounding environment (land use; height, type and functions of the buildings). The field observations were accompanied by detailed photographic documentation, thanks to which at stage III, it was possible to prepare exact diagrams of spatial development of individual sites.
- Semi-structured interviews, conducted with actors involved in urban agriculture, including with farm and garden owners, crop fieldworkers and vendors at local markets. They were carried out on the basis of the list of issues prepared during stage I. The semi-structured interview method allows for modification of the chronology of the questions asked, as well as for new issues to be raised both by the respondent and by the researcher (Longhurst 2003), thanks to which the scope of the information obtained was much broader than previously assumed. The conducted interviews allowed for collection of comprehensive information on the structural and production as well as organisational and technical characteristics of the analysed agricultural areas, urban farms and gardens. The interviews were not recorded and the collected information was entered in the questionnaire form either during the interviews or immediately after their completion.
- In-depth interviews with people in charge of managing urban agriculture in the selected cities – they were conducted with a representative of the Agri-Food and Veterinary Authority of Singapore, as well as a representative of the Kigali City Hall in charge of implementing the Kigali Master Plan 2013. The interviews provided information on the local and central authorities' policy towards the presence of urban agriculture within the space of selected cities as well as the related problems. Apart from that, in Kigali, an interview was conducted with Professor Rufus Jeyakumar, PhD, Dean of the School of Economics and Business Studies, Kigali Independent University. In Havana, conducting an interview with decision-makers in charge of urban agriculture management proved impossible. Instead, consultations were held with Professor Angelina Herrera Sorzano, a geographer working at the Faculty of Geography of the University of Havana (*Universidad de la Habana*), whose academic work is dedicated to urban agriculture in Cuba.

- Stage III (comparison and summary)

This stage consisted in analysis and synthesis of data collected during stages I and II and comparison of features and functions of urban agriculture in the analysed cities. It also included verification of the formulated research hypotheses as well as drawing up a number of cartography studies (diagrams and maps), presenting both distribution of agriculture within the designated research areas and the spatial organisation of individual farms, gardens and valleys used for agricultural purposes. David W. Harvey, one of the most frequently quoted socioeconomic geographers, proposes a number of types of academic explanation in his book on the methodology of geographical research, titled "Explanation in Geography" (1976). At stage III of developing this book, the following methods distinguished by D. Harvey were used:

- Morphometric analysis – which enables description of the spatial structure of the analysed phenomena. In the case of this book, it meant determination of the characteristics of distribution of urban agriculture within the analysed cities. On the basis of data collected during the field research and analysis of satellite images, a dynamic analysis of distribution of urban farms and gardens was also carried out. In Havana, the analysis covered the years 2000–2022, in Singapore 1974–2022 and in Kigali 2019–2022.
- Cause-and-effect analysis – enabling determination of the factors affecting distribution of urban agriculture in the selected cities, along with factors shaping its structural and production as well as organisational and technical characteristics.
- Functional analysis – enabling determination of the role of specific elements (in the case of this book, of urban agriculture) within larger structures (cities).

The last method applied in the book is comparative analysis of distribution and factors affecting the location of urban agriculture in the selected cities, along with its structural and production as well as organisational and technical characteristics. Thanks to the methods applied, it was possible to achieve the research objectives and to present a comprehensive view of urban agriculture in the three selected cities of the Global South.

The scope of use of the aforementioned research methods (in particular those included in stages I and II) differed slightly due to the different characteristics of the cities and of the urban agriculture within their space. Due to issues which arose during the fieldwork and the specific nature of research in individual cities, some of the methods had to be transformed and adapted to the local conditions. Moreover, after completion of each field research, the selected methods were verified and improved so as to make the research process more efficient and enable operationalisation of the stated research objectives. The scope of use of the research methods in individual cities was discussed in three chapters dedicated to each metropolis (Chapter 3 – Havana, Chapter 4 – Singapore and Chapter 5 – Kigali). It should also be emphasised that although in the case of Havana and Kigali, detailed research was limited to an arbitrarily designated, densely developed urban area, the reasoning applies to entire cities in all three instances.

1.5 Definition of urban agriculture[2]

There are multiple definitions of urban agriculture (also called urban farming) developed since the 1990s. The most compact and intuitive one, frequently quoted in literature on the subject (among others, by De Bon et al. 2010 or Zezza & Tasciotti

2010) is that proposed by R. Van Veenhuizen (2006, p. 2). According to it, urban agriculture means "the growing of plants and the raising of animals for food and other uses within and around cities and towns [...]". However, it should be emphasised that this definition was reproduced by the aforementioned authors in an abbreviated version. It continues as follows:

> [...] and related activities such as the production and delivery of inputs, and the processing and marketing of products. Urban Agriculture is located within or on the fringe of a city and comprises a variety of production systems, ranging from subsistence production and processing at household level to fully commercialised agriculture.

Therefore, this definition is much more comprehensive and includes a number of activities beyond its first part, which is the most frequently repeated in literature. A similar approach to urban agriculture was previously adopted by J. Smit et al. (2001) and L. J. A. Mougeot (2000), treating it as an industry (*sic!*) that includes both food production and its processing and distribution. On the other hand, H. de Zeeuw et al. (2011) emphasise that the definition of urban agriculture should be expanded to include transport, marketing as well as other non-agricultural services, such as agritourism.

Urban agriculture is a highly diversified activity in terms of production systems and can include both horticulture (including greenhouse farming, as well as modern hydroponic and aeroponic farming), aquaculture, animal husbandry, field cultivations typical of rural areas, and even agroforestry (Smit et al. 2001; Sroka 2014). In addition to that, agriculture can take various forms depending on the management methods and entities involved in the production process. Subsistence household gardens, community gardens, private commercial farms or state-owned agricultural farms are some of the many organisational forms of urban agriculture, all of which can occupy both private land and public spaces as well as wasteland zones, and sometimes even building rooftops (Altieri et al. 1999; Smit et al. 2001; Belevi & Baumgartner 2003). The entities involved in urban agriculture can be divided into three groups – state entities (national and local authorities involved in creating the policy pertaining to urban agriculture), non-state entities (units with international significance, such as FAO and international and local non-governmental organisations) and informal entities (Cissé et al. 2005, as quoted in Sroka 2014). The last group also includes individual agricultural producers (including natural persons and households, businesses and agricultural or horticultural establishments), which frequently form cooperatives, as well as providers of production resources who supply producers and vendors (Sroka 2014). Actors engaged in urban agriculture also include intermediaries, whose role goes beyond supplying vendors with products from producers. They can have a considerable impact on the development of prices as they are frequently the entities that achieve the biggest financial gains.

The aspect which gives rise to numerous doubts among urban agriculture researchers is the spatial dimension of urban agriculture. The expression "within or on the fringe of a city", used for instance by R. Van Veenhuizen (2006, p. 2), is insufficiently precise. L. J. A. Mougeot (2000), followed by numerous other authors (including De Bon et al. 2010; Pearson et al. 2010; De Zeeuw et al. 2011; Aubry et al. 2012), to a certain extent provide a more detailed definition by distinguishing between intra-urban agriculture, observed in densely developed urban areas (or within older, more established urban tissue), and peri-urban agriculture, located in the city outskirts. The same approach was also adopted in this book. Other authors tend to recognise any type of agricultural activity pursued within the administrative borders of a city as intra-urban agriculture, while activity outside those borders is considered peri-urban agriculture (Maxwell & Armar-Klemesu 1998; Moustier 1999). Nevertheless, it should be emphasised that distinguishing between intra-urban agriculture, peri-urban agriculture and rural agriculture is not always possible and typically depends on the local context and specificity of each analysed city, as well as the researchers themselves and their definition of "urban". Certain countries have introduced their own, official spatial coverage of urban agriculture, based on national and local provisions. For example, in Cuba, urban (both intra-urban and peri-urban) agriculture can be observed up to 10 km around the borders of provincial capitals (e.g. Havana, Cienfuegos, Santa Clara), up to 5 km around cities which are not provincial capitals and 2 km around smaller towns (Guevara Núñez 2001, as quoted in Díaz & Harris 2012). Certain authors, in turn, define the coverage of peri-urban agriculture on the basis of various criteria for the purposes of their own research. For instance, P. Moustier (1998), when referring to African cities, uses the maximum distance from which agricultural establishments are capable of supplying easily spoiling vegetables to the city centre. On the other hand, D. G. Mwamfupe (1994) sets the maximum distance that residents (in this case, Dar es Salaam in Tanzania) can travel on a daily basis to work in agriculture. In both cases, the spatial range urban agriculture will depend on how well the road and transport infrastructure is developed (Mougeot 2000). The following figures present examples of African intra-urban agriculture in Praia (Cabo Verde; see Figure 1.1) and peri-urban agriculture in Banjul (The Gambia; see Figure 1.2).

Certain definitions of urban agriculture, other than those based on spatial coverage and the aforementioned production systems, also include additional features that can be considered traits distinguishing it from rural agriculture. J. Smit et al. (2001) emphasise that urban agriculture is based on the use of intensive production methods, reuses natural resources and municipal waste and contributes to improved food security and well-being of residents, while simultaneously improving the quality of the natural environment. For this reason, urban agriculture, already at the level of its definition, is associated with the concept of sustainable development. S. Chaudhuri (2015) emphasises that the distinct feature of urban agriculture, setting it apart from rural agriculture, is its integration with the urban socioeconomic and ecological system. First, it supplies residents of cities with food and, at the same time, provides them with employment. Second, urban agriculture produces municipal waste and leads to its reuse through composting. Moreover, it participates in competition for spatial resources, which makes it a prominent element of the urban system. The multitude of functions fulfilled by urban agriculture merits taking a closer look at its role in the spatial and functional structure of cities.

Figure 1.1 Intra-urban agriculture in Praia, between the colonial district of *Platô* and one of the city's main thoroughfares – *Avenida Cidade de Lisboa.*

Source: Photograph by K. Górny, 2017.

Figure 1.2 Peri-urban agriculture in the vicinity of residential buildings in Bakau (within the metropolitan area of Banjul).

Source: Photograph by K. Górny, 2016.

1.6 The role of agriculture in spatial and functional structure of cities

As an integral part of a city's spatial and functional structure, agriculture plays an important role from the point of view of food producers and consumers as well as the urban ecosystem itself (Aubry et al. 2012). According to H. De Bon et al. (2010), one of the most distinct features of urban agriculture is its multifunctionality, defined as diverse economic, social and environmental roles assigned to it by the society. Therefore, the following presentation describes the main functions fulfilled by urban agriculture that are important from the perspective of residents (producers and consumers), as well as the urban ecosystem itself. The second part of the sub-chapter contains a description of key risks which improper management of agriculture within cities may entail.

1.6.1 Food security and sovereignty of the residents

As city populations increase, ensuring adequate services, infrastructure and especially access to food that meets the residents' needs and preferences becomes an increasingly difficult challenge. The most important and obvious advantage of urban agriculture is its contribution to food security, defined as a situation where the population has physical, social and economic access to a sufficient quantity of food corresponding to its nutritional needs and preferences (Jackson et al. 2016). There are three dimensions of food security listed in definitions of this concept that appear in literature. The first one is availability, that is, having a sufficient quantity of food to sustain a population's life. In other words, it is a sufficient food supply on the market. The second one is access, that is, physical (spatial) and economic access to food that corresponds to the population's preferences, while the third one – adequacy or utilisation, that is, a food ration that is balanced and safe for health (Ericksen 2008; Ingram 2011; Misselhorn et al. 2012). The stability of those three dimensions over time is particularly important. Apart from food security, another important term in research on sustainable development of urban areas is food sovereignty. This concept includes, in addition to unlimited access to food that corresponds to the population's needs and preferences, the right to decide on the method of food production and procurement, as well as determination of the level of self-sufficiency in this regard (Pimbert 2009). It is a term proposed during the FAO World Food Summit in 1996 by the international social movement associating farmers, *La Via Campesina*. Food sovereignty was supposed to supplement the commonly used concept of food security (Patel 2009).

Production being located within cities leads to increased food supply on the local market, supplementing, and in certain cases even serving as the basis, of diets of households, especially those with limited financial resources and low purchasing power (Bryld 2003; De Bon et al. 2010). Since urban agriculture is based on growing varied plant species and is less seasonal than rural agriculture, it can provide a better guarantee of food security, mainly due to the regularity of food supplies (De Bon et al. 2010). It plays a special role in areas called food deserts, where access to food is considerably limited (Thomaier et al. 2015). They are areas in a city without food stores or where existing food stores are financially inaccessible to the population. They are also described as zones with limited access to healthy, unprocessed food. Urban agriculture not only improves food security but also food sovereignty of residents, as it is frequently the result of grassroots initiatives of the local population. Therefore, residents can decide on their own how to produce and procure food.

1.6.2   Shortened supply chain

An important feature of urban agriculture is the fact that it operates within a shortened supply chain, both in spatial terms, understood as the distance between the place of production and place of distribution (and consumption), and in subjective terms, understood as the number of actors involved in the products' route. Since food is produced close to the markets, it does not require long-distance transport. This leads to reduction of transport costs and therefore the final price of the product. Fewer delivery vehicles entering the city, in turn, mean reduced fume emissions and therefore smog reduction (Specht et al. 2014). Urban farmers very often sell their products locally, which enables direct interactions between the producer and the consumer. Apart from economic benefits for both parties, resulting from reduced number of intermediaries (lower production costs for the producer and lower prices for the consumer), the shortened supply chain also allows for more control over the production process, frequently leading to sustained high quality of goods.

1.6.3   A source of income and employment

Urban agriculture, apart from contributing to improved food security of the residents and food self-sufficiency of cities, is also an important source of income, in particular for the poorest population. Farmers can produce food to meet their own needs and sell the surpluses, achieving higher and higher levels of production commercialisation. In both cases, households involved in urban agriculture save on food and can allocate the funds saved to other needs (Bryld 2003; Van Veenhuizen 2006). In cities where there is a shortage of jobs or competition on the labour market is high, agriculture is an important source of employment. However, it is frequently exclusively an additional gainful activity meant to improve the given family's financial standing. Income diversification is an important component of the strategy of securing one's livelihood, which can ensure financial security of a household during a crisis (Bryld 2003). What is more, due to the diversity of actors engaged in it, urban agriculture stimulates the development of the surroundings, that is, micro-enterprises involved in manufacturing production resources, such as seeds or fertilisers, food processing and packaging and finally also transport and sale.

1.6.4   Building inclusive and integrated urban communities

Urban agriculture also contributes to inclusion of excluded social groups, primarily the impoverished, disabled, elderly as well as immigrants, who – due to their social and financial situation – are on the margins of the labour market, struggling with securing the livelihood of their families (Van Veenhuizen 2006). Food production is therefore a strategy of reducing social disproportions, both in terms of access to food and other goods and in terms of the position on the labour market. This aspect is particularly important with respect to gender-based discrimination. In the majority of countries of the Global South, it is women who dominate among the people engaged in various stages of the food production process. Through involvement in farming, processing, transport, distribution and other activities, they gain a higher degree of independence and empowerment (Freeman 1993; Slater 2001; Hovorka 2006a, 2006b).

Other social functions fulfilled by urban agriculture should also be mentioned. Since it frequently takes the form of community gardens, taken care of by an organised group of residents, it plays an important role in building the society and strengthening connections between people. By involving multiple actors from various environments, it fosters development of a diverse network of connections, supporting the development of social capital, which is an important element of the strategy of securing one's livelihood. Residents involved in the operation of a community garden have an opportunity to exchange knowledge and experiences, reduce costs through shared production as well as obtain support in case of a crisis (Maldonado Villavicencio 2009; De Haan 2012).

Moreover, in many cities, there are gardens that fulfil educational functions. They are typically located next to schools, universities and other public institutions and are meant to provide education within the scope of ecological and organic food production and to raise awareness of the concept of sustainable development. Children living in the cities often have limited knowledge of how the products they consume are made because of their limited interactions with the countryside. Educational gardens tackle this problem by making city residents, especially the youngest ones, aware of how the food system they are a part of operates.

1.6.5 Environmental and landscape functions

Urban agriculture is an integral component of the urban ecosystem. When managed properly, it can help harmonise and regulate the processes that form parts of that ecosystem. Farming in urban areas entails a number of benefits for the natural environment. First of all, vegetation improves the air quality. Plants can absorb pollutants and produce oxygen, as well as trap dust in the soil. Crops also contribute to improvement of the microclimate by increasing humidity and reducing the air temperature, thus mitigating the urban heat island effect – which means lower costs of air conditioning of buildings during the summer period (Bryld 2003; Van Veenhuizen 2006). Urban agriculture also performs a retention role since the soil and the plants retain water and reduce surface runoff. This mitigates the negative effects of droughts and floods as well as takes the pressure off of the municipal sewage system (Bryld 2003; Aubry et al. 2012; Specht et al. 2014). The role played by urban agriculture in waste management also needs to be mentioned. Since sustainable and organic production techniques and methods are becoming increasingly popular, urban gardens commonly utilise compost. It is produced from waste generated within the gardens or farms, as well as wastes from nearby households and gastronomic points. Wastage of natural resources is also reduced thanks to application of production methods based on modern technologies, such as soilless cultivations, including hydroponics, aeroponics and aquaponics. They are based on closed or semi-closed systems, which minimise water, nutrient and energy losses. Their use is particularly beneficial in places with limited soil and water resources.

Urban agriculture frequently occupies urban wasteland zones as well as abandoned and neglected spaces, filling the so-called "gaps" in the urban tissue. It contributes to a more rational use of urban space, increases the share of greenery and therefore improves the landscape qualities of the city and the well-being of its residents (De Bon et al. 2010; Aubry et al. 2012).

1.6.6   *Risks associated with urban agriculture*

Agriculture in cities entails a number of environmental, economic and social benefits, as discussed above. However, potential risks and issues arising from improper management of agriculture and application of inappropriate agricultural techniques should also be examined. The first group of risks is associated with the use of animal faeces and waste from households to produce compost and fertilise the crops, as well as the use of untreated sewage for plant irrigation. This practice can lead to the spread of diseases such as cholera, typhoid or parasitic infections (e.g. tapeworm and hookworm) (Smit et al. 2001; Bryld 2003; Chaudhuri 2009; De Bon et al. 2010). Husbandry in the vicinity of residential housing frequently results in the spread of zoonoses, such as brucellosis and swine flu (Mougeot 2000), while an improperly secured compost bin emits harmful vapours, which can lead to bronchitis, dysentery or even cancers of the respiratory system (Bryld 2003). Lack of proper hygiene, including frequent hand washing and washing the fruit and vegetables picked, can contribute to transmission of diseases not only among producers, but also among consumers of the produced food. According to research by N. Chaudhuri (2009) carried out in Pikine in the metropolitan area of Dakar, use of untreated sewage in agriculture has caused many cases of parasitic infections, dysentery and other diseases. Farmers, unaware of the threat, have not been wearing proper protective clothing, exposing themselves to direct contact with sewage and solid waste. This example shows that without proper protective measures, amongst limited resources, urban agriculture can have a negative impact on the health of the farmers and/or gardeners as well as consumers of the food produced in urban areas.

There is also a number of studies in which urban agriculture is linked to the spread of malaria, among others, in Ghana, Tanzania and Cameroon (Afrane et al. 2004; Klinkenberg et al. 2008; Dongus et al. 2009; Stoler et al. 2009; Antonio-Nkondjio et al. 2011). Open water reservoirs within crop areas and fields as well as intensive irrigation support the spread of malaria-transmitting mosquitoes. Although more comprehensive research is needed in this regard, at the moment, the link between urban agriculture and development of this disease cannot be ruled out.

Another threat to the producers' health is poisoning caused by pesticides and other agrochemicals (e.g. mineral fertilisers) through their application without the use of proper safety measures, as well as by heavy metals, which are very frequent contaminants found in the soil in urban areas (Smit et al. 2001; Chaudhuri 2009). Toxic substances, including those present in the atmosphere, can be absorbed by plants, posing a threat to consumers of the food produced in cities (Smit et al. 2001; Maldonado Villavicencio 2009; De Bon et al. 2010; Redwood 2012). Leafy greens, which are the most common crop in urban agriculture, are particularly prone to pollutants.

Excessive use of fertilisers and pesticides, growing plants directly in the soil and in the vicinity of busy streets and industrial facilities, as well as inadequate waste management can lead to food contamination and pose a risk to health or even life of producers and consumers (Juarez 2009; Maldonado Villavicencio 2009). Other problems associated with practicing agriculture in urban areas include unpleasant odours and noise caused by farm animals, increased soil erosion due to the crops grown on steep slopes as well as spatial conflicts. Despite the listed risks, it should be emphasised that there are certain good practices which limit the negative aspects of agricultural activity in cities. Proper management and support for farmers' and gardeners' education and agricultural techniques used by them can almost completely eliminate the aforementioned risks. This way, urban agriculture can be a chance for improving the quality of life for city residents.

The issue of sustainable development of urban agriculture cannot be tackled without taking account of its multifunctionality, complexity and diverse links to other elements of the city's spatial and functional structure. Not only does it contribute to improved food security of the population, but also provides employment and an additional source of income. It even leads to social inclusion of excluded groups of residents. Apart from that, it fosters development of interpersonal bonds and fulfils a number of social or even cultural functions. In certain regions, for example, in Sub-Saharan Africa, urban agriculture is an integral part of cultural identity of the city residents. Therefore, maintaining it is a matter of preservation of local culture and protection of tradition (De Bon et al. 2010). With proper management, urban agriculture supports the balance of the city's ecosystem through reuse of organic waste, increased share of greenery in the total area and reduced transport of food from rural areas. In view of the risks associated with food production within city borders, it should be clearly stated that urban agriculture does not create benefits in an arbitrary manner. In order for its influence on the urban ecosystem and the life of residents to be positive, appropriate measures adapted to the local spatial, environmental, socioeconomic and even political conditions need to be applied. Poorly managed urban agriculture can lead to irreversible damage to the natural environment and pose a threat to the residents' health. In order to maximise the positive effects of local food production, it is necessary to include urban agriculture in spatial planning and long-term city development strategy (De Bon et al. 2010; Aubry et al. 2012). Integrated measures that include the municipal authorities, planners as well as residents can contribute to building sustainable food systems in the city, while simultaneously reducing the risk of aggravating local conflicts. Taking into account the fact that competition for space in urban areas is high, while agriculture rarely is a highly profitable activity, its future and role in the spatial and functional structure is frequently in the hands of decision-makers competent for spatial planning in cities (Aubry et al. 2012).

Notes

1   The economic crisis of 2007–2009 is also called the "3Fs crisis" – finance, fuel, food crisis – due to the fact that the collapse of the global financial system led to a drastic increase in fuel prices and therefore also food prices (Baker 2012; Chiripanhura & Niño-Zarazúa 2016).
2   This sub-chapter, as well as sub-chapter 1.6, uses fragments of an original chapter found in the following monograph: Górna, A. (2018). Urban agriculture: An opportunity for sustainable development. In *Globalización y desarrollo sostenible*, eds. M. Czerny, & C. A. Serna Mendoza. Warsaw: WUW, 129–142.

References

Afrane, Y. A., Klinkenberg, E., Drechsel, P., Owusu-Daaku, K., Garms, R., & Kruppa, T. (2004). Does irrigated urban agriculture influence the transmission of malaria in the city of Kumasi, Ghana? *Acta Tropica*, *89*(2), 125–134, https://doi.org/10.1016/j.actatropica.2003.06.001

Altieri, M. A., Companioni, N., Cañizares, K., Murphy, C., Rosset, P., Bourque, M., & Nicholls, C. I. (1999). The greening of the "barrios": Urban agriculture for food security in Cuba. *Agriculture and Human Values*, *16*(2), 131–140, https://doi.org/10.1023/a:1007545304561

Antonio-Nkondjio, C., Fossog, B. T., Ndo, C., Djantio, B. M., Togouet, S. Z., Awono-Ambene, P., & Ranson, H. (2011). Anopheles gambiae distribution and insecticide resistance in the cities of Douala and Yaoundé (Cameroon): Influence of urban agriculture and pollution. *Malaria Journal*, *10*(1), 154, https://doi.org/10.1186/1475-2875-10-154

Aubry, C., Ramamonjisoa, J., Dabat, M. H., Rakotoarisoa, J., Rakotondraibe, J., & Rabeharisoa, L. (2012). Urban agriculture and land use in cities: An approach with the multi-functionality and sustainability concepts in the case of Antananarivo (Madagascar). *Land Use Policy*, *29*(2), 429–439, https://doi.org/10.1016/j.landusepol.2011.08.009

Baker, J. L. (2012). *Directions in urban development. Impacts of financial, food and fuel crisis on the urban poor*. Urban Development Unit. The World Bank.

Belevi, H., & Baumgartner, B. (2003). A systematic overview of urban agriculture in developing countries from an environmental point of view. *International Journal of Environmental Technology and Management*, *3*(2), 193–211, https://doi.org/10.1504/ijetm.2003.003382

Boniface, P. (2003). *Atlas des relations internationales* [*Atlas of international relations*]. Paris: Armand Colin.

Boyd, A., & Comenetz, J. (2007). *An atlas of world affairs* (11th ed.). London: Routledge, https://doi.org/10.4324/9780203967522

Bryld, E. (2003). Potentials, problems, and policy implications for urban agriculture in developing countries. *Agriculture and Human Values*, *20*(1), 79–86, https://doi.org/10.1023/a:1022464607153

Chaudhuri, N. (2009). Using participatory education and action research for health risk reduction amongst farmers in Dakar, Senegal. In *Agriculture in urban planning generating livelihoods and food security*, ed. M. Redwood. London: IRDC, 181–199, https://doi.org/10.4324/9781849770439

Chaudhuri, S. (2015). Urban poor, economic opportunities and sustainable development through traditional knowledge and practices. *Global Bioethics*, *26*(2), 86–93, https://doi.org/10.1080/11287462.2015.1037141

Chiripanhura, B. M., & Niño-Zarazúa, M. (2016). The impacts of the food, fuel and financial crises on poor and vulnerable households in Nigeria: A retrospective approach to research inquiry. *Development Policy Review, 34*(6), 763–788, https://doi.org/10.1111/dpr.12183

Cissé, O., Gueye, N. F. D., & Sy, M. (2005). Institutional and legal aspects of urban agriculture in French-speaking West Africa: From marginalization to legitimization. *Environment and Urbanization, 17*(2), 143–154, https://doi.org/10.1177/095624780501700211

Czerny, M. (2012). *Bieda i bogactwo we współczesnym świecie: studia z geografii rozwoju* [*Poverty and wealth in the modern world: Studies in the geography of development*]. Wydawnictwa Uniwersytetu Warszawskiego.

De Bon, H., Parrot, L., & Moustier, P. (2010). Sustainable urban agriculture in developing countries. A review. *Agronomy for Sustainable Development, 30*(1), 21–32, https://doi.org/10.1051/agro:2008062

De Haan, L. J. (2012). The livelihood approach: A critical exploration. *Erdkunde*, 345–357, https://doi.org/10.3112/erdkunde.2012.04.05

De Zeeuw, H., Van Veenhuizen, R., & Dubbeling, M. (2011). The role of urban agriculture in building resilient cities in developing countries. *The Journal of Agricultural Science, 149*(S1), 153–163, https://doi.org/10.1017/s0021859610001279

Díaz, J. P., & Harris, P. (2012). Urban agriculture in Havana: Opportunities for the future. In *Continuous productive urban landscapes*, eds. A. Viljoen, & J. Howe. London: Routledge, 135–145, https://doi.org/10.4324/9780080454528-29

Dongus, S., Nyika, D., Kannady, K., Mtasiwa, D., Mshinda, H., Gosoniu, L., & Castro, M. C. (2009). Urban agriculture and Anopheles habitats in Dar es Salaam, Tanzania. *Geospatial Health, 3*(2), 189–210, https://doi.org/10.4081/gh.2009.220

Ericksen, P. J. (2008). Conceptualizing food systems for global environmental change research. *Global Environmental Change, 18*(1), 234–245, https://doi.org/10.1016/j.gloenvcha.2007.09.002

Forster, D., Buehler, Y., & Kellenberger, T. (2009). Mapping urban and peri-urban agriculture using high spatial resolution satellite data. *Journal of Applied Remote Sensing, 3*(1), 033523, https://doi.org/10.1117/1.3122364

Freedom House. (2023). *Freedom in the World 2023. Marking 50 years in the struggle for democracy.* Washington, DC: Author.

Freeman, D. B. (1993). Survival strategy or business training ground? The significance of urban agriculture for the advancement of women in African cities. *African Studies Review, 36*(3), 1–22, https://doi.org/10.2307/525171

Fücks, R. (2013). *Intelligent Wachsen. Die grüne Revolution* [*Smart growth: The green revolution*]. Carl Hanser Verlag GmbH & Co. KG, https://doi.org/10.3139/9783446434981

Górna, A. (2018). Urban agriculture: An opportunity for sustainable development. In *Globalización y desarrollo sostenible*, eds. M. Czerny, & C. A. Serna Mendoza. Warsaw: WUW, 129–142.

Górna, A., & Górny, K. (2020). Rolnictwo miejskie w miastach Afryki Subsaharyjskiej – ujęcie przestrzenne [Urban agriculture in Sub-Saharan African cities – spatial approach]. *Prace i Studia Geograficzne, 65*(4), 37–62.

Guevara Núñez, O. (2001). Demostración de que si se puede [Proof that you can]. In *Granma* (official newspaper of the Cuban Communist Party), February 1, 2001, p. 8.

Harvey, D. (1976). *Explanation in geography.* London: Edward Arnold.

Hovorka, A. J. (2006a). The No. 1 Ladies' Poultry Farm: A feminist political ecology of urban agriculture in Botswana. *Gender, Place & Culture, 13*(3), 207–225, https://doi.org/10.1080/09663690600700956

Hovorka, A. J. (2006b). Urban agriculture: Addressing practical and strategic gender needs. *Development in Practice, 16*(1), 51–61, https://doi.org/10.1080/09614520500450826

Ingram, J. (2011). A food systems approach to researching food security and its interactions with global environmental change. *Food Security, 3*, 417–431, https://doi.org/10.1007/s12571-011-0149-9

Jackson, P., Spiess, W. E., & Sultana, F. (2016). Introduction: Understanding the complexities of eating, drinking, and surviving. In *Eating, drinking: Surviving*, eds. P. Jackson, W. E. Spiess & F. Sultana. Cham: Springer Nature, 1–12, https://doi.org/10.1007/978-3-319-42468-2

Jantakat, Y., Juntakut, P., Plaiklang, S., Arree, W., & Jantakat, C. (2019). Spatiotemporal change of urban agriculture using google earth imagery: A case of municipality of Nakhonratchasima City, Thailand. *The International Archives of the Photogrammetry, Remote Sensing and Spatial Information Sciences, 42*, 1301–1306, https://doi.org/10.5194/isprs-archives-xlii-2-w13-1301-2019

Juarez, H. (2009). Water contamination and its impact on vegetable production in the Rimac River, Peru. In *Agriculture in urban planning generating livelihoods and food security*, ed. M. Redwood. London: IRDC, 125–143, https://doi.org/10.4324/9781849770439

Klinkenberg, E., McCall, P. J., Wilson, M. D., Amerasinghe, F. P., & Donnelly, M. J. (2008). Impact of urban agriculture on malaria vectors in Accra, Ghana. *Malaria Journal, 7*(1), 1–9, https://doi.org/10.1186/1475-2875-7-151

Longhurst, R. (2003). Semi-structured interviews and focus groups. *Key Methods in Geography, 3*(2), 143–156.

Maldonado Villavicencio, L. (2009). Urban agriculture as a livelihood strategy in Lima, Peru. In *Agriculture in urban planning generating livelihoods and food security*, ed. M. Redwood. London: IRDC, 49–70, https://doi.org/10.4324/9781849770439

Malthus, T. R. (1798). *An essay on the principle of population*. London: J. Johnson.

Maxwell, D., & Armar-Klemesu, M. (1998). *Urban agriculture: Introduction and review of literature. Accra*. Noguchi Memorial Institute for Medical Research.

Meadows, D. H., Randers, J., & Meadows, D. L. (2013). *The limits to growth (1972)*. New Haven: Yale University Press, 101–116, https://doi.org/10.12987/9780300188479-012

Misselhorn, A., Aggarwal, P., Ericksen, P., Gregory, P., Horn-Phathanothai, L., Ingram, J., & Wiebe, K. (2012). A vision for attaining food security. *Current Opinion in Environmental Sustainability, 4*(1), 7–17, https://doi.org/10.1016/j.cosust.2012.01.008

Mougeot, L. J. (2000). *Urban agriculture: Definition, presence, potentials and risks, and policy challenges*. Cities feeding people series. London: IDRC.

Moustier, P. (1998). La complémentarité entre agriculture urbaine et agriculture rurale [The complementarity between urban agriculture and rural agriculture]. In *Agriculture urbaine en Afrique de l'Ouest: une contribution à la sécurité alimentaire et à l'assainissement des villes*, ed. O. B. Smith. Wageningen: CTA/Ottawa: IDRC, 41–55.

Moustier, P. (1999). *Filières maraîchères à Brazzaville: quantification et observatoire pour l'action [Vegetable sectors in Brazzaville: Quantification and observatory for action]*. Montpellier: CIRAD – Agrisud International – Agricongo.

Mwamfupe, D. G. (1994). *Changes in agricultural land use in the peri-urban zone of Dar es Salaam, Tanzania* (Doctoral dissertation, ProQuest Dissertations & Theses).

Nouschi, M. (2016). *Petit Atlas historique du XXe siècle [Small historical atlas of the 20th century]*. Paris: Armand Colin.

Patel, R. (2009). Food sovereignty. *The Journal of Peasant Studies, 36*(3), 663–706, https://doi.org/10.1080/03066150903143079

Pearson, L. J., Pearson, L., & Pearson, C. J. (2010). Sustainable urban agriculture: Stocktake and opportunities. *International Journal of Agricultural Sustainability, 8*(1–2), 7–19, https://doi.org/10.3763/ijas.2009.0468

Pimbert, M. (2009). *Towards food sovereignty*. London: International Institute for Environment and Development.

Pulighe, G., & Lupia, F. (2016). Mapping spatial patterns of urban agriculture in Rome (Italy) using Google Earth and web-mapping services. *Land Use Policy, 59*, 49–58, https://doi.org/10.1016/j.landusepol.2016.08.001

Redwood, M. (2012). *Introduction to agriculture in urban planning generating livelihoods and food security*. London: IRDC, 1–20, https://doi.org/10.4324/9781849770439

Slater, R. J. (2001). Urban agriculture, gender and empowerment: An alternative view. *Development Southern Africa, 18*(5), 635–650, https://doi.org/10.1080/03768350120097478

Smit, J., Nasr, J., & Ratta, A. (2001). *Urban agriculture: Food, jobs and sustainable cities*. New York: Urban Agriculture Network, http://www.jacsmit.com/book.html

Solarz, M. W. (2012). North–South, Commemorating the First Brandt Report: Searching for the contemporary spatial picture of the global rift. *Third World Quarterly, 33*(3), 559–569, https://doi.org/10.1080/01436597.2012.657493

Solarz, M. W. (2014). *The language of global development: A misleading geography*. London: Routledge, https://doi.org/10.4324/9780203077382

Solarz, M. W. (2019). *The global North-South atlas: Mapping global change*. London: Routledge, https://doi.org/10.4324/9780429492037

Specht, K., Siebert, R., Hartmann, I., Freisinger, U. B., Sawicka, M., Werner, A., & Dierich, A. (2014). Urban agriculture of the future: an overview of sustainability aspects of food production in and on buildings. *Agriculture and Human Values, 31*(1), 33–51, https://doi.org/10.1007/s10460-013-9448-4

Sroka, W. (2014). Definicje oraz formy miejskiej agrokultury–przyczynek do dyskusji. *Wieś i Rolnictwo, 164*(3), 85–103.

Stoler, J., Weeks, J. R., Getis, A., & Hill, A. G. (2009). Distance threshold for the effect of urban agriculture on elevated self-reported malaria prevalence in Accra, Ghana. *The American Journal of Tropical Medicine and Hygiene, 80*(4), 547–554, https://doi.org/10.4269/ajtmh.2009.80.547

Szymańska, D., & Korolko, M. (2015). *Inteligentne miasta: idea, koncepcje i wdrożenia* [*Smart cities: idea, concepts and implementation*]. Toruń: Wydawnictwo Naukowe UMK.

Taylor, J. R., & Lovell, S. T. (2014). Urban home food gardens in the Global North: Research traditions and future directions. *Agriculture and Human Values, 31*(2), 285–305, https://doi.org/10.1007/s10460-013-9475-1

Thomaier, S., Specht, K., Henckel, D., Dierich, A., Siebert, R., Freisinger, U. B., & Sawicka, M. (2015). Farming in and on urban buildings: Present practice and specific novelties of Zero-Acreage Farming (ZFarming). *Renewable Agriculture and Food Systems, 30*(1), 43–54, https://doi.org/10.1017/s1742170514000143

Un-Habitat. (2011). *Cities and climate change: Global report on human settlements, 2011*. London: Routledge, https://doi.org/10.4324/9781849776936

United Nations Department of Economic and Social Affairs. (2019). *World urbanization prospects: The 2018 revision*. New York: Author, https://doi.org/10.18356/b9e995fe-en

United Nations Development Programme. (2022). *Human development report 2021/2022. Uncertain times, unsettled lives: Shaping our future in a transforming world*. New York: Author, https://doi.org/10.18356/9789210016407c003

van Veenhuizen, R. (2006). *Introduction to Cities farming for the future: Urban agriculture for green and productive cities*, ed. R. Van Veenhuisen René. RUAF Foundation, IIRR, IDRC, Ottawa, Canada, 1–17, https://idrc-crdi.ca/en/book/cities-farming-future-urban-agriculture-green-and-productive-cities

Zezza, A., & Tasciotti, L. (2010). Urban agriculture, poverty, and food security: Empirical evidence from a sample of developing countries. *Food Policy, 35*(4), 265–273, https://doi.org/10.1016/j.foodpol.2010.04.007

2 Contemporary research on urban agriculture

In the face of intensive urbanisation and spatial expansion of urban buildings, climate change and challenges to food security of city residents, urban agriculture is a prominent subject touched upon in research in multiple academic disciplines, including agronomics, sociology, economics, urban studies as well as geography. Studies on urban agriculture in various regions of the world have different themes, approaches to the subject matter of the research, as well as research methods applied. In the case of those dedicated to countries of the Global South, discussions on the subject of food security and sovereignty prevail, along with the role played by local food production in securing the livelihood of city residents. With respect to metropolises of the Global North, it is much more common to discuss issues related with improving self-sufficiency of cities, as well as to describe the role played by local food production in improving the condition of the natural environment and building social connections. This chapter aims to outline the contemporary directions of research on urban agriculture, which differ depending on the socioeconomic and political conditions in a given region. The first part of this chapter will discuss urban agriculture in cities of the Global South, while the second part – in cities of the Global North. Confrontation of the dominant trends, research narratives as well as research results that are different for both of the aforementioned regions is meant to present urban agriculture as a subject matter of research in the broadest possible context.

Any list of exhaustive monographs that serve as a comprehensive introduction to the issue of urban agriculture would not be complete without the works of L. J. A. Mougeot (2000) and J. Smit et al. (2001). They concern the issues of definitions, spatial scope and role of urban agriculture within the spatial and functional structure of cities around the world. Both publications are widely quoted and serve as theoretical academic handbooks on research on urban agriculture. Their authors are considered leading researchers on this subject matter. Luc J. A. Mougeot, who is associated with the International Development Research Centre (IDRC) – a Canadian public institution – has initiated many research grants for young scholars conducting research on urban agriculture in various regions of the world. Jac Smit, on the other hand, was a pioneer in research on urban agriculture, as well as a proponent of its protection. Because of it, other authors and collaborators have called him "the father of urban agriculture" (Jac Smit 2021).

DOI: 10.4324/9781003429845-2

Apart from the listed theoretical items, there are many noteworthy works based on cases studies, which provide a review of diverse issues associated with food production within cities. They are based on the experience of the authors conducting research in various regions of the world. They include the book titled *Agropolis: The social, political, and environmental dimensions of urban agriculture*, published in 2005 and edited by the aforementioned L. J. A. Mougeot. *Agropolis*, supplemented with the editor's theoretical preface, covers a number of case studies concerning the issue of supplying cities with food, the role of women in the process of food production or food security of the most impoverished city residents. Other prominent reviews include two monographs published by the aforementioned IDRC – (1) "Cities farming for future, Urban Agriculture for green and productive cities" (2006), edited by R. Van Veenhuizen, prepared in collaboration with The RUAF Global Partnership on Sustainable Urban Agriculture and Food Systems, and (2) "For hunger-proof cities: Sustainable urban food systems" (1999), edited by M. Koc, R. MacRae, L. J. A. Mougeot and J. Welsh, as well as "Growing cities, growing food: urban agriculture on the policy agenda. A reader on urban agriculture", edited by N. Bakker, M. Dubbeling, S. Guendel, U. Sabel-Koschella and H. De Zeeuw, published in 2000 by the German Foundation for International Development (GFID). Studies edited by A. Viljoen (2005), in which a team of researchers present the concept of Continuous Productive Urban Landscapes (CPUL) based on varied case studies, are also particularly important from the point of view of spatial research. The concept is based on the assumption that so-called productive spaces, that is, spaces taken up, for example, by urban agriculture, should be included in urban planning so that they can become an integral component of the urban substance. The publication describes a number of solutions aligned with this concept, which are being or can be implemented in contemporary cities. A new version of the book published by Routledge in 2014 (Viljoen & Bohn 2014) evaluates the activities within the scope of using the CPUL proposal in various locations.

2.1 Urban agriculture in cities of the Global South

Many countries of the Global South struggle with the issue of unequal access to food. It is caused by unfair distribution of agricultural crops, domination of crops intended for export, weakness of the national institutions, corruption, technological underdevelopment as well as demographic boom and ecosystem changes. Continuously expanding cities are therefore facing numerous problems resulting from the need to supply the ever-growing population with food. For this reason, research on urban agriculture in countries of the Global South is strictly connected with issues of food security, which has been the subject of a number of analyses (Maxwell 1995; Armar-Klemesu 2000; Bryld 2003; Zezza & Tasciotti 2010; Crush et al. 2011; Mkwambisi et al. 2011; Redwood 2012; Gallaher et al. 2013; Lynch et al. 2013; Badami & Ramankutty 2015). Some authors also point out the role of local food production in building sustainable and resilient cities (Smit & Nasr 1992; Hamm & Baron 1999; Deelstra & Girardet 2000; Smit et al. 2001; Dubbeling & Merzthal 2006; Pearson et al. 2010; Spiaggi 2010; De Zeeuw et al. 2011; Giradet 2005;

Górna 2018). A number of studies also analyse the multifunctionality of urban agriculture (Ba 2008; Henderson 2009; Yang et al. 2010; Aubry et al. 2012). Other identified roles of agriculture, apart from its role in supplying city residents with food and improving their food security, include environmental, social or recreational functions. Z. Yang et al. (2010) and J. C. Henderson (2009), on the basis of research carried out in Beijing and Singapore, respectively, also point out the potential for using urban or peri-urban farms in tourism. Particularly noteworthy similarities can be found in conclusions drawn by two authors, A. Ba and I. Zasada, conducting research in completely different socioeconomic conditions in two regions, the Global South and the Global North, indicating the opportunities entailed by the multifunctionality of agriculture in cities. A. Ba (2008), on the basis of research carried out in the peripheries of Dakar, demonstrates that agriculture in the capital city of Senegal can hardly be considered multifunctional nowadays; however, introduction of new functions could present a chance for its maintenance in the future. On the other hand, I. Zasada (2011), in reference to the peripheries of cities in Western Europe, emphasises that diversification of the activity of urban farms and multifunctional approach to development strengthens it and may contribute to agriculture remaining on a highly competitive market. The conclusions of both authors are particularly important in the context of the case study discussed in this book – Singapore, where urban farms are quite common.

Another prominent review work is an article by A. J. Hamilton et al. (2014), which identifies the need for research on six issues concerning urban agriculture in cities in developing countries. These issues are: the share of population engaged in urban agriculture on a global and regional scale, the relationship between urban agriculture and spread of diseases as well as its role in combating malnutrition and obesity, the impact of urban agriculture on women's empowerment, methods of obtaining institutional support for urban agriculture as well as the risks presented by chemical pollution in food production. According to the author, it is those problems that require the fastest reaction on the part of the academic circle.

Due to the diverse features of urban agriculture, resulting from local environmental, political and especially socioeconomic conditions, the predominant trends in research are presented below, divided into three regions: Latin America, South, East and Southeast Asia and Sub-Saharan Africa. Apart from the results of analyses carried out by scholars from various centres, they also demonstrate the key problems faced by people engaged in and managing urban agriculture. It should be emphasised that researchers from the region of the Global South play a very distinct role among those involved in the subject matter. In addition to providing a unique perspective and knowledge of the specific nature of the analysed regions, they are also key when it comes to establishing the directions of future research.

2.1.1   *Latin America*

The most frequently discussed examples of agriculture in cities of Latin America come from Cuba. This country is an excellent illustration of how quickly and dynamically urban agriculture can develop in an economic crisis. As a global-scale phenomenon, Havana and other cities on the biggest island in the Caribbean are described by many authors in relation to the concept of sustainable development, food security and sovereignty, as well as building resilient cities (see Rodríguez Castellón 2003; Koont 2007, 2008, 2011; Herrera Sorzano 2009, 2015; Nelson et al. 2009; Simón Reardon & Pérez 2010; Gürcan 2014; Leitgeb et al. 2016). Havana was the main focus in analyses developed in the second half of the 1990s and at the beginning of this century. It was also at that time when agriculture, which had initially appeared within the city space as a grassroots movement by the residents, received state support, thanks to which it started developing on a larger and larger scale. Comprehensive studies dedicated to urban agriculture in the Cuban capital city include especially articles by M. A. Altieri et al. (1999), S. G. Chaplowe (1998) and S. Koont (2009), based on previously conducted field research, as well as texts by C. Murphy (1999) and M. G. Novo & C. Murphy (2000), which contain an issue-based presentation of the genesis and specificity of urban agriculture in the capital city of Cuba with respect to its food self-sufficiency and food security of its residents.

Other important items include works by the anthropologist A. Premat (2003, 2005, 2009), in which she analyses small urban gardens in Havana from a sociocultural perspective. In 2012, A. Premat also published a book titled *Sowing Change*, which contains a comprehensive discussion of development of urban gardens (primarily small-scale ones) in Havana and confronts the situation of urban gardeners with the authorities' aspirations when it comes to developing sustainable agriculture (Premat 2012). In her works, A. Premat demonstrates that gardens in Havana are more than just production spaces. They are simultaneously places where deeper bonds between residents are developed and an integrated local community emerges. O. Plonska (2017) reaches similar conclusions. Furthermore, in her 2003 work, A. Premat proves the thesis proposed by Henri Lefebvre (1968), according to which space can be used as a means of control, domination and power. Urban gardens in Havana are dependent on the authorities' decisions regarding their location as well as forms of operation.

Other particularly noteworthy studies are those based on the aforementioned CPUL concept – using the example of Havana (Díaz & Harris 2005) as well as Cienfuegos (Viljoen & Howe 2005). The work by A. Viljoen and J. Howe is one of few studies which tackle the spatial dimension of urban agriculture. The authors point out that due to small building density, gardens are also located in central, colonial districts. What determines the distribution of urban agriculture in Cienfuegos are therefore the features of its spatial structure. Another Cuban city analysed in literature on the subject is Trinidad. In her work combining sociology and political ecology, C. Buchmann (2009) demonstrates that the functioning of urban gardens contributes to building networks of mutual assistance and support among the residents of Trinidad, which contributes to increasing the social capital of households and improving their resilience to crisis situations. Such conclusions are particularly important in the context of the role that urban agriculture can play in contemporary as well as future environmental, political, economic and social crises.

Other important works in international literature are those by A. H. Herrera Sorzano – a Cuban urban agriculture researcher. In addition to publications dedicated to the role of urban gardens in ensuring food sovereignty of the residents of Cuban cities (2009, 2015), another noteworthy work is the geographic atlas developed by a Cuban team of scholars and edited by Herrera Sorzano: "*Artemisa: atlas agrícola de una provincia cubana*" (2016), which presents agriculture (including urban agriculture) in the Cuban province of Artemisa.

The Latin American studies literature on the subject also includes many analyses of Mexico City and peri-urban agriculture in its metropolitan area. An important issue taken up by numerous authors here is the relationship between urban and rural areas, as well as the issue of disappearance of arable land in favour of sprawling residential building zones of the Mexican capital city (Losada et al. 1998; Arias Luján 2000; Galván & Duque 2000; Lima et al. 2000; Losada et al. 2000; Galindo & Delgado 2006; Sánchez 2009). The aforementioned authors base their research on the livelihood approach. According to this approach, every household has five types of livelihoods assets, which determine its level of livelihood security. These are: natural capital (natural resources and environmental services), financial capital (economic assets, which include cash, savings as well as basic infrastructure and equipment), human capital (the household members' skills, knowledge and ability to work), social capital (social networks, relations and interpersonal bonds) as well as physical capital (technical infrastructure and production resources) (Scoones 1998; De Haan 2012). Urban agriculture is an important component of the strategy of securing one's livelihood. It contributes to ensuring food security of the population, is an additional source of income and fosters deeper interpersonal bonds, thus improving the social capital.[1] Research on peri-urban agriculture in Mexico was also carried out in Toluca, where it concerned primarily traditional maize cultivation in the metropolitan area (Lerner & Appendini 2011; Lerner et al. 2013; Lerner et al. 2014). The authors demonstrate that building expansion leads to changes in the role of urban agriculture. Although it is being eliminated, it is not disappearing completely from that space. However, it is turning primarily into subsistence agriculture. The population grows maize to meet its own needs and only sells small surpluses. Profits from agriculture are merely an additional source of income and a method of protection against a crisis since the majority of residents are employed in other sectors of the economy. The authors also indicate the importance of the support from the government and urban agriculture being taken into account in spatial planning for the analysed households.

Another noteworthy city in Latin America is Rosario, Argentina (see works by Dubbeling & Merzthal 2006; Propersi 2008; Ponce & Donoso 2009; Spiaggi 2010; Dubbeling 2015; Piacentini et al. 2014; Battiston et al. 2017), as an example of a metropolis where sustainable agriculture is promoted and monitored by the authorities. It was included in the city's development strategy, among others, due to the serious threat to small farms in the peripheries posed by single-crop farming of soy intended for export. Rosario is an example of a city where the authorities' support has a positive impact on the functioning of urban agriculture.

Other subjects analysed in literature on urban agriculture in Latin America include those that fit in with the gender studies movement. The research concerns mainly the impact of gender on the roles fulfilled by men and women in urban agriculture. What is more, international literature frequently tackles the issues of women's empowerment as a result of involvement in food production. Research within the scope of gender studies has been conducted, among others, in Magdalena, Mexico (Buechler 2009) and Lima (Soto et al. 2006). The results demonstrate important differences in the role of men and women in the process of food production and distribution. Women are typically involved in subsistence farming since their role is to ensure the food security of households. Men, on the other hand, typically perform tasks associated with processing and sales. Moreover, they are engaged in commercial agriculture more frequently than women since they are responsible for providing financial resources to households.

Research on cities of Latin America is dominated by analyses of urban agriculture as an important element of the strategy of securing one's livelihood. Case studies from the region, those concerning both cities and peripheral areas, indicate that urban agriculture contributes to improving the food security of residents, is an additional source of income for them and has a positive impact on building interpersonal connections. What is more, it is also a form of securing households against a crisis.

2.1.2   South, East and Southeast Asia

Taking into account the fact that the majority of cities in South, East and Southeast Asia are undergoing intensive development and experiencing the urban sprawl phenomenon, urban agriculture, especially in peripheral locations, is subject to significant building pressure. Farms located in the outskirts of large metropolises of the region are shut down or removed to more and more remote locations and replaced by new housing estates. Many authors tackle issues associated with the endangered position of urban agriculture in city outskirts as its spatial coverage changes, as well as the relations between urbanisation and food production in peri-urban areas. Noteworthy case studies in literature include primarily numerous works on Hanoi. These are, among others, works by R. B. Thapa (2003), R. B. Thapa et al. (2004), R. B. Thapa and Y. Murayama (2008) and V. C. Pham et al. (2015), based on spatial analysis of satellite images. The authors analyse the pace of sprawl of the buildings into peri-urban agricultural areas, the land use breakdown and agricultural usability of soils within the city. The results of their research demonstrate the effectiveness of automated remote sensing methods in research on peri-urban agriculture. Apart from that, the authors point out the need for integrated measures within the scope of spatial planning in order to protect agricultural areas and mitigate the negative effects of urban sprawl, such as development and fragmentation of crop fields leading to reduction in food supplies to the city and deterioration of the residents' food security. Urban agriculture in the capital city of Vietnam, studied using automated remote sensing methods, was also analysed by D. Forster et al. (2009). On the other hand, N. P. Le and N. M. Dung (2018) conducted research on multifunctionality of peri-urban farms in Hanoi. The results of this research demonstrate that although agriculture is not the main source of income of the households they analysed, it plays an important role in securing the livelihood of the population. It contributes to improved food security and is a source of additional income as well as employment for people who find it difficult to find work outside the agricultural sector. What is more, peri-urban farms, in particular those involved in growing decorative plants, are a tourist attraction.

Agriculture in Chinese cities is also worthy of note. The text by Z. Feifei et al. (2009) emphasises the role of agriculture in shaping local communities of migrants from urban areas, who settle in the outskirts of Chinese metropolises. On the other hand, C. Zhong et al. (2020) point to the relationship between urbanisation and development of urban agriculture by demonstrating that the increase in city population of China is positively correlated with the value added of urban agriculture per resident. Z. Yang et al. (2010), based on research carried out in the peripheries of Beijing, indicate that development of agritourism presents a chance for preserving agriculture within the city space and improving the efficiency of land use. The authors emphasise that in Chinese cities, it is becoming increasingly frequent to establish farms that combine production and tourist functions, thus diversifying the sources of income.

Many authors also tackle the issue of peri-urban agriculture in Indian cities. For instance, research at the level of households in which urban agriculture is an important component of the strategy of securing their livelihood was carried out by U. R. Gowda et al. (2012) in Bangalore and Z. Hussain and M. Hanisch (2014) in Hyderabad. At this point, it should also be mentioned that there is a number of works dedicated to the relationship between agriculture and food security in various cities of Southeast Asia: Chennai (Nambi et al. 2014) and Indian metropolises in general (Marshall & Randhawa 2017), Hanoi (Pulliat 2015), Jakarta (Chandra & Diehl 2019) as well as Beijing (Du et al. 2012).

Due to the proximity of a large market and culinary preferences of the residents, the area around Asian cities is dominated primarily by vegetable production (Midmore & Jansen 2003); hence, vegetable cultivation and supply to urban areas in various geographic contexts is an important issue. Works dedicated to this subject matter include, among others, an article on Ho Chi Minh City by H. G. P. Jansen et al. (1996). The authors of the publication emphasise that the share of vegetables in the production structure in the peripheries of the metropolis has been growing successively since the 1980s as a result of market transformations in the country and increased demand for vegetables among city residents. At present, vegetable cultivation, which is much more profitable, is replacing cultivation of other food crops such as rice in Vietnamese cities (Khai et al. 2007). Similar transformations also took place in the 1980s in Hong Kong (Lau 2013).

In recent years, there has also been increased interest in placement of crops on building rooftops in Asian cities. The potential of development of rooftop agriculture based on the hydroponic method was an issue tackled, among others, with respect to Guangzhou (Liu et al. 2016; Su et al. 2020). Su et al. propose a bold thesis that the building rooftop surface in the city is sufficiently large to achieve complete self-sufficiency in terms of vegetable production. Similar research demonstrating the potential of rooftop farming and the benefits it entails for the urban ecosystem and city residents was carried out in Hong Kong (Hui 2011; Wang & Pryor 2019), George

Town, Malaysia (Hashim et al. 2018), Dhaka (Safayet et al. 2017) as well as Singapore, which is analysed in this book (Astee & Kishnani 2010; Diehl et al. 2020).

Due to the high share of fish and seafood in Asian diet (Yeung 1987), aquaculture is also described in literature on the subject. It was analysed, among others, in Dhaka (Islam et al. 2004), Chattogram (Hossain et al. 2009), Hanoi (Hoan & Edwards 2005) and Bangkok (Mrozik et al. 2019).

2.1.3   Sub-Saharan Africa

Urban agriculture began developing dynamically in Sub-Saharan Africa in the 1980s. It was then that crops became an integral element of the landscape of growing cities in the region. At the beginning of the decade, only 15%–20% of the population throughout the whole African continent was involved in urban agriculture. On the other hand, already in early 1990s, this percentage was approaching 70% (Bryld 2003). For example, in Dar es Salaam, the number of households engaged in urban agriculture increased from 18% in 1967 to 67% in 1991 (Ratta & Nasr 1996). The reason for such a drastic change throughout the continent was primarily intensified migration from rural areas to cities caused by demographic boom. E. Bryld (2003) also emphasises that an important factor forcing city residents to become involved in local food production was serious deterioration of their economic situation as a result of implementation of structural adjustment projects promoted by the World Bank and the International Monetary Fund under the agreement concluded in Bretton Woods. They were meant to support African countries in the process of market liberalisation and restructuring of agriculture. However, the implemented programmes led to deeper indebtedness and crisis of the first sector of the economy in the majority of the countries on the continent (Bello 2009). The deteriorated situation of the farmers, reduction of employment in agriculture and serious reduction of crops resulted in limited food supplies. The urban population was affected the most by those food shortages. As a result, urban agriculture became the response to the serious consequences of the food crisis in the 1980s that affected the city residents in the region. Increased engagement in agriculture in cities of Sub-Saharan Africa also results from their dynamic spatial expansion. As a result of the building sprawl process, vast, typically rural, agricultural areas were included within the newly set out administrative boundaries. The new urban farmers are therefore former rural farmers, who have been "absorbed" by the spatially developing city.

The described research on urban agriculture in African cities is vast and thematically diverse. Many publications concern the contribution of urban agriculture to improved food security of the residents, fight against malnutrition and reduction of poverty. Some authors conducted their analyses at the level of households, based on the aforementioned livelihood approach. Those include D. G. Maxwell (1995) and D. G. Maxwell et al. (1998) in Kampala, M. Armar-Klemesu and D. G. Maxwell (1998) in Accra, E. Njogu (2012) in Nairobi, as well as N. Speybroeck et al. (2004) in Brazzaville. Their research demonstrates that agriculture is key to securing the livelihood of the analysed households. The food produced can be allocated to meeting the basic nutritional needs of household members or sold in order to obtain additional income. The results of research on the role of urban agriculture in combating malnutrition, carried out by D. G. Maxwell et al. (1998) and A. Mboganie-Mwangi and D. W. J. Foeken (1996), are different. D. G. Maxwell et al. indicate that agriculture in Kampala contributes to improved nutrition of children under 5 years of age. However, such impact was not demonstrated in the case of Nairobi (Mboganie-Mwangi & Foeken 1996). Different conclusions concerning the role of local food production in empowerment of and obtaining independence by women were included in works by D. G. Maxwell (1995) and M. Armar-Klemensu and D. G. Maxwell (2000). The study by D. G. Maxwell (1995) dedicated to Kampala demonstrates that urban agriculture is a way for women (who make up the great majority of people involved in agricultural production) to become financially independent from their husbands. On the other hand, the study by M. Armar-Klemesu and D. G. Maxwell (1998) on Accra demonstrated that the role of women in urban agriculture is much smaller than is the case in Kampala. In the capital city of Ghana, the people engaged in agricultural production are predominantly men, while women, in order to obtain their own income and achieve a certain level of financial independence, seek employment outside agriculture. This shows that depending on cultural and institutional factors, the role of urban agriculture in women's empowerment can vary. Studies dedicated to how local production contributes to improving food security and reducing poverty also include those on Harare (Drakakis-Smith et al. 1995; Mutonodzo 2009) as well as Lilongwe and Blantyre (Mkwambisi et al. 2011).

Research on urban agriculture in Sub-Saharan African countries highlights the risks created by poorly managed urban agriculture practiced amongst limited resources. B. F. Tano et al. (2011) indicate the danger posed to the health of producers and consumers by excessive use of artificial fertilisers and use of untreated municipal sewage to water the crops in Yamoussoukro. Problems associated with improper plant irrigation in Addis Ababa, on the other hand, were studied by D. J. Van Rooijen et al. (2010). To water the crops, residents of the city use water from the Akaki River, which is where most of untreated sewage from households and industrial plants is released. This practice creates the risk of soil contamination, and therefore also of contamination of food plants, among others, with heavy metals, whose excessive concentration in the river and in vegetables produced in the city has also been observed by other authors (including Itanna 2002; Chary et al. 2008). Similar conclusions have also been drawn by A. Etale and D. C. Drake (2013) with respect to Kigali. The authors point to the risk posed by crops being concentrated in the bottom of the Nyabugogo river valley, where untreated sewage from industrial plants is released. Other authors analysing the issue of improper crop irrigation that can lead to contamination of soil and food produced within the city include Karanja et al. (2009), who studied Nairobi, and Srikanth, Naik (2004), who analysed Asmara. One should also mention the work by M. Lydecker and P. Drechsel (2010) on Accra. The authors present a different perspective on the issue of municipal sewage being used in agriculture, treating it as a chance for improving and relieving the municipal sewage system.

Figure 2.1 Lettuce cultivation in Bissau. In the distance, there is a fence separating the crop field from a wild landfill.

Source: Photograph by A. Górna, February 2020.

On the other hand, G. Nabulo et al. (2012), on the basis of research on agriculture in Kampala, drew attention to an issue associated with soil contamination resulting from plots of land previously used to store municipal and industrial waste being allocated to agriculture. Similar observations, concerning food crops being adjacent to informal landfills, were made by the author of this book during field research carried out in Bissau in February 2020 (see Figures 2.1 and 2.2).

An important issue connected with agriculture in Sub-Saharan African countries is its connection to the spread of diseases. For instance, many authors have analysed the relationship between the location of urban agriculture and spread of malaria since presence of agricultural land, and therefore standing water, within the city creates advantageous breeding grounds for mosquitoes, as demonstrated by researchers. Research on this subject was carried out inter alia by Y. A. Afrane et al. (2004) in Kumasi (Ghana), E. Klinkenberg et al. (2008) and J. Stoler et al. (2009) in Accra, S. Dongus et al. (2009) in Dar es Salaam, as well as C. Antonio-Nkondjio et al. (2011) in Douala and Yaoundé.

Another important subject taken up in the context of African countries, which fits in with the gender studies movement, is the role of men and women in urban agriculture and ensuring food security of households. *Women feeding cities: Mainstreaming gender in Urban Agriculture and Food Security* (2009), edited by A. Hovorka, H. De Zeeuw and M. Njenga, is an important example of this. The publication contains case studies of different cities on the continent, including Accra

Figure 2.2 A landfill where waste is regularly incinerated, located in the immediate vicinity
of crops.

Source: Photograph by A. Górna, February 2020.

(Hope et al. 2009), Kampala (Nabulo et al. 2009), Harare (Toriro 2009), Kisumu
(Ishani 2009) and Nairobi (Kuria Gathuru et al. 2009) or Pikine (Gaye & Touré
2009). Most of those studies are by scholars from African countries. A. Hovorka
has also tackled the aforementioned subject in other works on cities in Botswana
(including 2005, 2006). The author emphasised that restructuring of urban agriculture (the texts talk about peri-urban agriculture) can present a chance for women to
renegotiate their marginalised position. However, it is important for women to be
able to determine their role in the food production process themselves. Otherwise,
their involvement in agriculture can be a burden, not a method of empowerment.

Finally, one should also mention the monograph edited by G. Prain, N. Karanja
and D. Lee-Smith (2010) – *African Urban Harvest: Agriculture in the Cities of
Cameroon, Kenya and Uganda*, which is a collection of case studies on urban agriculture in three countries – Cameroon, Kenya and Uganda. The publication, based
on the livelihood concept, also includes works pertaining to food security, building sustainable food systems as well as the policy of municipal authorities with
respect to local agricultural production. They describe experience within the scope
of managing urban agriculture in different socioeconomic, institutional and political conditions, which is meant to contribute to improving its efficiency. The work
was dedicated to the memory of Jac Smit – the precursor of research on urban
agriculture in African cities.

2.2 Urban agriculture in cities of the Global North

Compared with cities of the Global South, in the case of cities of the Global North, the issue of food security of the residents is tackled much less frequently. In highly developed countries, supplying the urban population with basic food is not as serious of a problem as it is in less developed countries. However, considerable economic disparities within the bloc of countries of the Global North cannot be omitted. Due to low purchasing power of the most impoverished social groups and continuously increasing prices of food products, urban agriculture frequently presents a chance for all citizens to achieve full food security.

Since one of the distinct features of urban agriculture is that it starts being practiced primarily in times of crisis, urban gardens were established in various cities when the food security of the residents was at risk (Mok et al. 2014; Burgin 2018). In the United States, such gardens already appeared during World War I and were referred to as war gardens. Afterwards, they became common during the great crisis of the 1930s (when they were known as relief gardens) and during World War II (so-called victory gardens) (Mok et al. 2014). The gardens were meant to supply residents with food and, especially during the interwar period, provide employment to a large number of unemployed Americans. After World War II, local food production ceased to be that significant, primarily due to common use of refrigerators, introduction of supermarkets and popularity of the consumer lifestyle. H. F. Mok et al. (2014) point out that a revival or urban gardens in American cities could be observed at the turn of the 1960s and the 1970s, when the first social pro-ecological movements emerged.

Similar developments took place in European countries. During both world wars, due to limited food supplies, urban agriculture, primarily in the form of small self-supply gardens, allotment gardens and community gardens played an important role in urban food systems (Mok et al. 2014; Burgin 2018). After World War II, in turn, the agricultural policy of countries of Western Europe was focused on producing large quantities of inexpensive food in order to prevent famine and improve availability of food products. Many countries decided to invest in large-scale, capital-intensive agriculture (Deelstra et al. 2001). On the other hand, as European cities expanded, more and more agricultural land was transformed into housing estates. Despite the fact that increasing building density inside cities and development of post-industrial areas have been observed in cities on the continent, spatial expansion is responsible for the disappearance of urban agriculture to a significant extent (Zasada 2011). T. Deelstra et al. (2001) point out that at the end of the 20th century, emphasis shifted to the potential benefits of local food production and proximity of the market. Urban agriculture, in particular agriculture that fulfils a number of social functions, was more and more frequently considered an advantageous element of cities' spatial and functional structure.

Contemporary research on urban agriculture in countries of the Global North tackles a number of subjects. First, many academic texts concern the issue of urban agriculture's potential in building sustainable, green cities. The book *Urban*

Agriculture Europe, published in 2016 and edited by F. Lohrberg, L. Lička, L. Scazzosi and A. Timpe, is a particularly important collective work discussing this subject matter. It was prepared with participation of several dozen European authors, thanks to which it contains a number of case studies from various countries on the continent, including Spain, Germany, Italy, Bulgaria, Switzerland, Ireland as well as Poland. Other prominent works are those by L. M. Boukharaeva and M. Marloie (2006), S. Burgin (2018) and B. Gulyas and J. Edmondson (2021).

Taking into account the fact that social and cultural significance of local food production is being taken note of with increased frequency, while many urban gardens and farms are diversifying their activity by including recreational, training and educational or even tourist services, some authors focus on multifunctionality of urban agriculture in Europe (Deelstra et al. 2001; Zasada 2011), the United States (Lovell 2010; Poulsen et al. 2017) and Canada (Valley & Wittman 2019). Among non-agricultural functions of peri-urban agriculture, which can gain prominence in view of its post-production development, I. Zasada (2011) lists, inter alia, its environmental functions (which are rarely appreciated by city residents, in the researcher's opinion), recreational, landscape, educational and social functions. According to I. Zasada, traditional functions and values of agriculture have been replaced by new ones, not related to food production and more focused on food consumption and recreational activities. On the other hand, W. Valley and H. Wittman (2019), based on research dedicated to urban farms in Vancouver, emphasised the environmental function of agriculture, including primarily the reduction of so-called food miles, that is, the distance travelled by food products from their place of production to their place of consumption; participation in building a multifunctional urban landscape, as well as the role in increasing so-called food literacy or agricultural literacy, that is, awareness and knowledge (including practical knowledge) of residents on food production, distribution and consumption, as well as composting. Apart from the above functions, M. N. Poulsen et al. (2017) also mentions – with respect to community gardens in Baltimore – the role of urban agriculture in improving the perception of neighbourhoods among their residents, which aligns more broadly with the urban beautification movement.

Important subjects in research on urban agriculture of the region are also allotment gardens, which were primarily typical of European cities. A particularly noteworthy work is the book edited by S. Bell et al. (2016) titled *Urban Allotment Gardens in Europe*, which comprehensively outlines the subject matter of allotment gardens while taking into account their legal standing and location within the urban tissue, as well as ecosystem services they provide and social functions they fulfil. The book is dominated by examples from European cities. Other works which analyse the issue of allotment gardens include studies by I. Cabral et al. (2017) and J. Breuste and M. Artmann (2015), which analyse their ecosystem services in Salzburg and Leipzig, respectively. Allotment gardens also remain popular to this day in cities of Eastern Europe, for example, in Polish cities (see, for instance, works dedicated to the Tri-City – Moskalonek 2020, Wrocław, Katowice and Krakow – Klepacki, Kujawska 2018 or Poznań – Poniży, Stachura 2017). They have already been common since World War I (Bellows 2004; Trembecka &

Figure 2.3 Prinzessinnengarten in Berlin, located in a former cemetery.
Source: Photograph by K. Górny, September 2019.

Kwartnik-Pruc 2018) and significantly gained importance after the end of World War II, during the period of the Polish People's Republic.

In recent years, something of a renaissance of community gardens has been observed in European and North American cities. Social, educational and recreational functions play a much more important role than the production function. Those gardens can be described as run by the community to meet the community's needs and an important effect of their activity is strengthening of interpersonal bonds as well as building a robust group of socially active residents. Cities where community garden movements have been developing intensively in recent years include Berlin, Paris, Warsaw and New York. Figures 2.3 and 2.4 present community gardens operating in Berlin and Letchworth Garden City.

Since community gardens are exceptionally popular in American cities, two books with reviews dedicated to them should be mentioned among studies on the subject. The first one is *City Bountiful: A Century of Community Gardening in America* by L. J. Lawson, published in 2005 and concerning the history of development of community gardens in the United States, while the other one is *Community Gardening as Social Action* (2016) by C. Nettle, which presents this subject matter in the context of Australian cities.

The latest examples of literature on the subject, dedicated to community and allotment gardens, based on case studies of various cities of the Global North, also deserve attention. These are, for example, works on Brighton and Hove in Great Britain (Nicholls et al. 2020), Lugo in Spain (Gómez-Villarino & Ruiz-Garcia 2021), Adelaide in Australia (Hume et al. 2021), Portland and Vancouver

Figure 2.4 The Wynd Show Garden in Letchworth Garden City.
Source: Photograph by K. Górny, July 2018.

(McClintock et al. 2021), as well as metropolitan areas of London, Berlin, Milan and Rotterdam (Zasada et al. 2019).

Although the issue of food security of residents in the Global North is not as pressing as in the Global South, I. Opitz et al. (2016) emphasise that the subject is becoming increasingly important, in particular with respect to cities where access to healthy, safe and varied food corresponding to the needs and financial capabilities of residents is becoming more and more limited. An example of a metropolis where serious food shortages have been observed and the share of so-called food deserts has increased is Detroit, Michigan. The city, facing more and more exacerbated social and economic problems, which resulted in a decla-ration of bankruptcy in 2013, saw development of so-called agrihoods (a blend of two words: agriculture and neighbourhoods) – neighbourhoods which integrate the residential function with urban agriculture, with an active, organised com-munity of residents. The case of Detroit is another example of urban agriculture appearing within the city space amongst limited food supplies and food security being at risk. The subject of development of urban gardens in Detroit has been taken up by several authors, including M. M. White (2011), K. J. Colasanti et al. (2012) and E. Giorda (2012).

In cities of the Global North, similarly to cities of Southeast Asia, modern urban agriculture using cutting-edge technology is becoming more and more prominent. Some examples of research on application of innovative methods and solutions within the scope of urban agriculture include works concerning so-called ZFarming or Zero-Acreage Farming, that is, agriculture based on minimising the space used, including rooftop farms and farms located inside buildings (Specht et al. 2014; Thomaier et al. 2015; Specht et al. 2016).

The COVID-19 pandemic, announced by the World Health Organisation in 2020, highlighted the existing problems associated with lack of food security and once again emphasised the importance of local food production. Similar risks associated with a sudden interruption of global supply chains were demonstrated by the war in Ukraine and grain shipments to African countries which depend on them being withheld. According to R. Lal (2020), urban agriculture, in cities of the Global South as well as Global North, can serve as an idea for resolving serious economic crises, which result in limited availability of affordable food products. It contributes to improved food security, more varied diet of city residents, as well as reduced use of natural resources. Therefore, one can anticipate that urban agriculture will be more and more frequently discussed in academic works, including in the context of cities of the Global North.

Note

1 Other research at the household level using the livelihood concept in cities of Latin America has been conducted, among others, in Lima (Niñez 1985; Dasso & Pinzas 2000; Maldonado Villavicencio 2012), Belém (Madaleno 2000), Managui (Shillington 2013) and La Paz (Kreinecker 2000).

References

Afrane, Y. A., Klinkenberg, E., Drechsel, P., Owusu-Daaku, K., Garms, R., & Kruppa, T. (2004). Does irrigated urban agriculture influence the transmission of malaria in the city of Kumasi, Ghana? *Acta Tropica*, *89*(2), 125–134, https://doi.org/10.1016/j. actatropica.2003.06.001

Altieri, M. A., Companioni, N., Cañizares, K., Murphy, C., Rosset, P., Bourque, M., & Nicholls, C. I. (1999). The greening of the "barrios": Urban agriculture for food security in Cuba. *Agriculture and Human Values*, *16*(2), 131–140, https://doi.org/10.1023/a: 1007545304561

Antonio-Nkondjio, C., Fossog, B. T., Ndo, C., Djantio, B. M., Togouet, S. Z., Awono-Ambene, P., & Ranson, H. (2011). Anopheles gambiae distribution and insecticide resistance in the cities of Douala and Yaoundé (Cameroon): Influence of urban agriculture and pollution. *Malaria Journal*, *10*(1), 154, https://doi.org/10.1186/1475-2875-10-154

Arias Luján, E. (2000). Aprovechamiento de la agricultura urbana en la Ciudad de México [Taking advantage of urban agriculture in Mexico City]. In *Procesos metropolitanos y agricultura urbana*, ed. T. Lima, Unidad Xochimilco: Universidad Autónoma Metropolitana, 239–242.

Armar-Klemesu, M. (2000). Urban agriculture and food security, nutrition and health. In *Growing cities, growing food: Urban agriculture on the policy agenda. A reader on urban agriculture*, eds. N. Bakker, M. Dubbeling, S. Guendel, U. Sabel-Koschella, & H. De Zeeuw, Deutsche Stiftung fur Internationale Entwicklung (DSE), 99–117, https://ruaf.org/ document/growing-cities-growing-food/

Astee, L. Y., & Kishnani, N. T. (2010). Building integrated agriculture: Utilising rooftops for sustainable food crop cultivation in Singapore. *Journal of Green Building*, *5*(2), 105–113, https://doi.org/10.3992/jgb.5.2.105

Aubry, C., Ramamonjisoa, J., Dabat, M. H., Rakotoarisoa, J., Rakotondraibe, J., & Rabeharisoa, L. (2012). Urban agriculture and land use in cities: An approach with the multifunctionality and sustainability concepts in the case of Antananarivo (Madagascar). *Land Use Policy*, *29*(2), 429–439, https://doi.org/10.1016/j.landusepol.2011.08.009

Ba, A. (2008). L'agriculture de Dakar: quelle multifonctionnalité et quelles perspectives [Agriculture in Dakar: what multifunctionality and what prospects]. In *La diversité de l'agri-culture urbaine dans le monde vol. 3*, ed. R. Vidal. Ecole nationale supérieure du paysage, Université Paris Nanterre, 43–54.

Badami, M. G., & Ramankutty, N. (2015). Urban agriculture and food security: A critique based on an assessment of urban land constraints. *Global Food Security*, *4*, 8–15, https://doi.org/10.1016/j.gfs.2014.10.003

Bakker, N., Dubbeling, M., Guendel, S., Sabel-Koschella, U., & Zeeuw, H. D. (2000). *Growing cities, growing food: Urban agriculture on the policy agenda. A reader on urban agriculture*. Deutsche Stiftung fur Internationale Entwicklung (DSE), Zentralstelle fur Ernahrung und Landwirtschaft, https://ruaf.org/document/growing-cities-growing-food/

Battiston, A., Porzio, G., Budai, N., Martínez, N., Pérez Casella, Y., Terrile, R., & Paz, N. (2017). Green Belt Project: Promoting agroecological food production in peri-urban Rosario. *Urban Agriculture Magazine*, *33*, 52–54.

Bell, S., Fox-Kämper, R., Keshavarz, N., Benson, M., Caputo, S., Noori, S., & Voigt, A. (2016). *Urban allotment gardens in Europe*. London: Routledge, https://doi.org/10.4324/9781315686608

Bello, W. (2009). *The food wars*. London: Verso Books.

Bellows, A. C. (2004). One hundred years of allotment gardens in Poland. *Food & Foodways*, *12*(4), 247–276, https://doi.org/10.1080/07409710490893793

Boukharaeva, L. M., & Marloie, M. (2006). Family urban agriculture as a component of human sustainable development. *CAB Reviews: Perspectives in Agriculture, Veterinary Science, Nutrition and Natural Resources*, *1*(025), 1–10, https://doi.org/10.1079/pavsnnr2006025

Breuste, J. H., & Artmann, M. (2015). Allotment gardens contribute to urban ecosystem service: Case study Salzburg, Austria. *Journal of Urban Planning and Development*, *141*(3), https://doi.org/10.1061/(asce)up.1943-5444.0000264

Bryld, E. (2003). Potentials, problems, and policy implications for urban agriculture in developing countries. *Agriculture and Human Values*, *20*(1), 79–86, https://doi.org/10.1023/a:1022464607153

Buchmann, C. (2009). Cuban home gardens and their role in social–ecological resilience. *Human Ecology*, *37*(6), 705–721, https://doi.org/10.1007/s10745-009-9283-9

Buechler, S. (2009). Gender dynamics of fruit and vegetable production and processing in peri-urban Magdalena, Sonora, Mexico.). In *Women feeding cities: Mainstreaming gender in urban agriculture and food security*, eds. A. Hovorka, H. De Zeeuw, & M. Njenga. CTA/Practical Action, 181–198, https://doi.org/10.3362/9781780440460.012

Burgin, S. (2018). 'Back to the future'? Urban backyards and food self-sufficiency. *Land Use Policy*, *78*, 29–35, https://doi.org/10.1016/j.landusepol.2018.06.012

Cabral, I., Keim, J., Engelmann, R., Kraemer, R., Siebert, J., & Bonn, A. (2017). Ecosystem services of allotment and community gardens: A Leipzig, Germany case study. *Urban Forestry & Urban Greening*, *23*, 44–53, https://doi.org/10.1016/j.ufug.2017.02.008

Chandra, A. J., & Diehl, J. A. (2019). Urban agriculture, food security, and development policies in Jakarta: A case study of farming communities at Kalideres–Cengkareng district, West Jakarta. *Land Use Policy, 89*, 104211, 1–11, https://doi.org/10.1016/j. landusepol.2019.104211

Chaplowe, S. G. (1998). Havana's popular gardens: Sustainable prospects for urban agriculture. *Environmentalist, 18*(1), 47–57, https://doi.org/10.1023/a:1006582201985

Chary, N. S., Kamala, C. T., & Raj, D. S. S. (2008). Assessing risk of heavy metals from consuming food grown on sewage irrigated soils and food chain transfer. *Ecotoxicology and Environmental Safety, 69*(3), 513–524, https://doi.org/10.1016/j.ecoenv.2007.04.013

Colasanti, K. J., Hamm, M. W., & Litjens, C. M. (2012). The city as an "agricultural powerhouse"? Perspectives on expanding urban agriculture from Detroit, Michigan. *Urban Geography, 33*(3), 348–369, https://doi.org/10.2747/0272-3638.33.3.348

Crush, J., Hovorka, A., & Tevera, D. (2011). Food security in Southern African cities: The place of urban agriculture. *Progress in Development Studies, 11*(4), 285–305, https://doi.org/10.1177/146499341001100402

Dasso, A., & Pinzas, T. (2000). NGO experiences in Lima targeting urban poor through urban agriculture. In *Growing cities, growing food: Urban agriculture on the policy agenda. A reader on urban agriculture*, eds. N. Bakker, M. Dubbeling, S. Guendel, U. Sabel-Koschella, & H. De Zeeuw. Deutsche Stiftung fur Internationale Entwicklung (DSE), Zentralstelle fur Ernahrung und Landwirtschaft, 349–361, https://ruaf.org/document/growing-cities-growing-food/

De Haan, L. J. (2012). The livelihood approach: A critical exploration. *Erdkunde*, 345–357, https://doi.org/10.3112/erdkunde.2012.04.05

De Zeeuw, H., Van Veenhuizen, R., & Dubbeling, M. (2011). The role of urban agriculture in building resilient cities in developing countries. *The Journal of Agricultural Science, 149*(S1), 153–163, https://doi.org/10.1017/s0021859610001279

Deelstra, T., & Girardet, H. (2000). Urban agriculture and sustainable cities. In *Growing cities, growing food. Urban agriculture on the policy agenda*, eds. N. Bakker, M. Dubbeling, S. Gündel, U. Sabel-Koshella, & H. de Zeeuw. Feldafing, Germany: Zentralstelle für Ernährung und Landwirtschaft (ZEL), 43–66, https://ruaf.org/document/growing-cities-growing-food/

Deelstra, T., Boyd, D., & Van Den Biggelaar, M. (2001). Multifunctional land use: An opportunity for promoting urban agriculture in Europe. *Urban Agriculture Magazine, 4*, 33–35.

Díaz, J. P., & Harris, P. (2005). Urban agriculture in Havana: Opportunities for the future. In *Continuous productive urban landscapes*, eds. A. Viljoen, J. Howe. London: Routledge, 135–145, https://doi.org/10.4324/9780080454528-29

Diehl, J. A., Sweeney, E., Wong, B., Sia, C. S., Yao, H., & Prabhudesai, M. (2020). Feeding cities: Singapore's approach to land use planning for urban agriculture. *Global Food Security, 26*, 100377, https://doi.org/10.1016/j.gfs.2020.100377

Dongus, S., Nyika, D., Kannady, K., Mtasiwa, D., Mshinda, H., Gosoniu, L., & Castro, M. C. (2009). Urban agriculture and Anopheles habitats in Dar es Salaam, Tanzania. *Geospatial Health, 3*(2), 189–210, https://doi.org/10.4081/gh.2009.220

Drakakis-Smith, D., Bowyer-Bower, T., & Tevera, D. (1995). Urban poverty and urban agriculture: An overview of the linkages in Harare. *Habitat International, 19*(2), 183–193, https://doi.org/10.1016/0197-3975(94)00065-a

Du, S. S., Cai, J. M., Guo, H., & Fan, Z. W. (2012). Food security-oriented urban agriculture development typologies: A case study of vegetable production in peri-urban Beijing. *Progress in Geography, 31*(6), 783–791.

Dubbeling, M. (2015). *Integrating urban agriculture and forestry into climate change action plans*. RUAF Foundation, 1–18.

Dubbeling, M., & Merzthal, G. (2006). Sustaining urban agriculture requires the involvement of multiple stakeholders. In *Cities farming for the future. Urban agriculture for green and productive cities*, ed. R. Van Veenhuizen. RUAF Foundation, the Netherlands, IDRC, Canada and IIRR publishers, the Philippines, 19–40, https://idrc-crdi.ca/en/book/cities-farming-future-urban-agriculture-green-and-productive-cities

Etale, A., & Drake, D. C. (2013). Industrial pollution and food safety in Kigali, Rwanda. *International Journal of Environmental Research, 7*(2), 403–406.

Feifei, Z., Jianming, C., & Gang, L. (2009). How urban agriculture is reshaping peri-urban Beijing? *Open House International, 34*(2), 15–24, https://doi.org/10.1108/ohi-02-2009-b0003

Forster, D., Buehler, Y., & Kellenberger, T. (2009). Mapping urban and peri-urban agriculture using high spatial resolution satellite data. *Journal of Applied Remote Sensing, 3*(1), 033523, https://doi.org/10.1117/1.3122364

Galindo, C., & Delgado, J. (2006). Los espacios emergentes de la dinámica rural-urbana. *Problemas del desarrollo, 37*(147), 187–216, https://doi.org/10.22201/iiec.20078951e.2006.147.7639

Gallaher, C. M., Kerr, J. M., Njenga, M., Karanja, N. K., & WinklerPrins, A. M. (2013). Urban agriculture, social capital, and food security in the Kibera slums of Nairobi, Kenya. *Agriculture and Human Values, 30*(3), 389–404, https://doi.org/10.1007/s10460-013-9425-y

Galván, F., & Duque, A. (2000). Producción urbana de alimentos y microempresas sociales en el Distrito Federal [Urban food production and social microenterprises in the Federal District]. In *Procesos metropolitanos y agricultura urbana*, ed. T. Lima, 231–238.

Gaye, G., & Touré, M. N. (2009). Gender and urban agriculture in Pikine, Senegal. In *Women feeding cities: Mainstreaming gender in urban agriculture and food security*, eds. A. Hovorka, H. De Zeeuw, & M. Njenga. CTA/Practical Action, 219–233, https://doi.org/10.3362/9781780440460.014

Giorda, E. (2012). Farming in Mowtown: Competing narratives for urban development and urban agriculture in Detroit. In *Sustainable food planning: Evolving theory and practice,* eds. A. Viljoen, & J. S. Wiskerke, Wageningen: Wageningen Academic Publishers, 269–279, https://doi.org/10.3920/978-90-8686-187-3_23

Giradet, H. (2005). Urban agriculture and sustainable urban development. In *Continuous productive urban landscapes*, eds. A. Viljoen & J. Howe. London: Routledge, 51–58, https://doi.org/10.4324/9780080454528-15

Gómez-Villarino, M. T., & Ruiz-Garcia, L. (2021). Adaptive design model for the integration of urban agriculture in the sustainable development of cities. A case study in northern Spain. *Sustainable Cities and Society, 65*, 102595, https://doi.org/10.1016/j.scs.2020.102595

Górna, A. (2018). Urban agriculture: An opportunity for sustainable development. In *Globalización y desarollo sostenible*, eds. M. Czerny, & C. A. Serna Mendoza, 129–142.

Gowda, U. R., Chandrakanth, M. G., Srikanthamurthy, P. S., Yadav, C. G., Nagaraj, N., & Channaveer. (2012). Economics of peri-urban agriculture: Case of Magadi Off Bangalore. *Economic and Political Weekly, 47*(24), 75–80.

Gulyas, B. Z., & Edmondson, J. L. (2021). Increasing city resilience through urban agriculture: Challenges and solutions in the Global North. *Sustainability, 13*(3), 1465, https://doi.org/10.3390/su13031465

Gürcan, E. C. (2014). Cuban agriculture and food sovereignty: Beyond civil-society-centric and globalist paradigms. *Latin American Perspectives, 41*(4), 129–146, https://doi.org/10.1177/0094582x13518750

Hamilton, A. J., Burry, K., Mok, H. F., Barker, S. F., Grove, J. R., & Williamson, V. G. (2014). Give peas a chance? Urban agriculture in developing countries. A review. *Agronomy for Sustainable Development*, *34*(1), 45–73, https://doi.org/10.1007/s13593-013-0155-8

Hamm, M. W., & Baron, M. (1999). Developing an integrated, sustainable urban food system: The case of New Jersey, United States. In *For hunger-proof cities: Sustainable urban food systems*, eds. M. Koc, R. MacRae, L. J. A. Mougeot, & J. Welsh. IDRC, 54–59, https://idrc-crdi.ca/en/book/hunger-proof-cities-sustainable-urban-food-systems

Hashim, N. H., Mohd Hussain, N. H., & Ismail, A. (2018). The rise of rooftop urban farming at George Town, Penang. *Environment-Behaviour Proceedings Journal*, *3*(7), 351–355, https://doi.org/10.21834/e-bpj.v3i7.1242

Henderson, J. C. (2009). Agro-tourism in unlikely destinations: A study of Singapore. *Managing Leisure*, *14*(4), 258–268, https://doi.org/10.1080/13606710903204456

Herrera Sorzano, A. (2009). Impacto de la agricultura urbana en Cuba [Impact of urban agriculture in Cuba]. *Novedades en Población*, *5*(9), 1–14.

Herrera Sorzano, A. H. (2015). La soberanía alimentaria desde la agricultura urbana: un reto para el desarrollo de la producción de alimentos en Cuba [Food sovereignty from urban agriculture: A challenge to the development of food production in Cuba]. *Revista GeoNordeste*, 1, 150–172.

Hoan, V. Q., & Edwards, P. (2005). Wastewater reuse through urban aquaculture in Hanoi, Vietnam: Status and prospects. *Urban aquaculture. CABI International, Wallingford*, 103–117, https://doi.org/10.1079/9780851998299.0103

Hope, L., Cofie, O., Keraita, B., & Drechsel, P. (2009). Gender and urban agriculture: The case of Accra, Ghana. In *Women feeding cities: Mainstreaming gender in urban agriculture and food security*, eds. A. Hovorka, H. De Zeeuw, & M. Njenga. CTA/Practical Action, 65–78, https://doi.org/10.3362/9781780440460.004

Hossain, M. S., Chowdhury, S. R., Das, N. G., Sharifuzzaman, S. M., & Sultana, A. (2009). Integration of GIS and multicriteria decision analysis for urban aquaculture development in Bangladesh. *Landscape and Urban Planning*, *90*(3–4), 119–133, https://doi.org/10.1016/j.landurbplan.2008.10.020

Hovorka, A. J. (2005). The (re) production of gendered positionality in Botswana's commercial urban agriculture sector. *Annals of the Association of American Geographers*, *95*(2), 294–313, https://doi.org/10.1111/j.1467-8306.2005.00461.x

Hovorka, A. J. (2006). The No. 1 Ladies' Poultry Farm: A feminist political ecology of urban agriculture in Botswana. *Gender, Place & Culture*, *13*(3), 207–225, https://doi.org/10.1080/09663690600700956

Hovorka, A., Zeeuw, H. D., & Njenga, M. (2009). *Women feeding cities: Mainstreaming gender in urban agriculture and food security*. CTA/Practical Action, https://doi.org/10.3362/9781780440460.000

Hui, S. C. (2011). Green roof urban farming for buildings in high-density urban cities. In *Hainan China world green roof conference 2011*, 1–9.

Hume, I. V., Summers, D. M., & Cavagnaro, T. R. (2021). Self-sufficiency through urban agriculture: Nice idea or plausible reality? *Sustainable Cities and Society*, *68*, 102770, https://doi.org/10.1016/j.scs.2021.102770

Hussain, Z., & Hanisch, M. (2014). Dynamics of peri-urban agricultural development and farmers' adaptive behaviour in the emerging megacity of Hyderabad, India. *Journal of Environmental Planning and Management*, *57*(4), 495–515, https://doi.org/10.1080/09640568.2012.751018

Ishani, Z. (2009). Key gender issues in urban livestock keeping and food security in Kisumu, Kenya. *Women feeding cities: Mainstreaming gender in urban agriculture and food security*, eds. A. Hovorka, H. De Zeeuw, & M. Njenga. CTA/Practical Action, 106–121, https://doi.org/10.3362/9781780440460.007

Islam, M. S., Chowdhury, M. T. H., Rahman, M. M., & Hossain, M. A. (2004). Urban and peri-urban aquaculture as an immediate source of food fish: Perspectives of Dhaka City, Bangladesh. *Urban Ecosystems*, *7*(4), 341–359, https://doi.org/10.1007/s11252-005-6834-8

Itanna, F. (2002). Metals in leafy vegetables grown in Addis Ababa and toxicological implications. *Ethiopian Journal of Health Development*, *16*(3), 295–302, https://doi.org/10.4314/ejhd.v16i3.9797

Jac Smit. (2021). *Welcome to 21st century agriculture*, http://www.jacsmit.com/ (accessed 6.06.2021).

Jansen, H. G. P., Midmore, D. J., Binh, P. H., Valasayya, S., & Tru, L. C. (1996). Profitability and sustainability of peri-urban vegetable production systems in Vietnam. *NJAS Wageningen Journal of Life Sciences*, *44*(2), 125–143, https://doi.org/10.18174/njas.v44i2.552

Karanja, N. N., Njenga, M., Prain, G., Kangâ, E., Kironchi, G., Githuku, C., & Mutua, G. K. (2009). Assessment of environmental and public health hazards in wastewater used for urban agriculture in Nairobi, Kenya. *Tropical and Subtropical Agroecosystems*, *12*(1), 85–97.

Khai, N. M., Ha, P. Q., & Öborn, I. (2007). Nutrient flows in small-scale peri-urban vegetable farming systems in Southeast Asia – a case study in Hanoi. *Agriculture, Ecosystems & Environment*, *122*(2), 192–202, https://doi.org/10.1016/j.agee.2007.01.003

Klinkenberg, E., McCall, P. J., Wilson, M. D., Amerasinghe, F. P., & Donnelly, M. J. (2008). Impact of urban agriculture on malaria vectors in Accra, Ghana. *Malaria Journal*, *7*(1), 1–9, https://doi.org/10.1186/1475-2875-7-151

Koc, M., MacRae, R., Mougeot, L. J., & Welsh, J. (1999). *For hunger-proof cities: Sustainable urban food systems*. IDRC, Ottawa, ON, CA, https://idrc-crdi.ca/en/book/hunger-proof-cities-sustainable-urban-food-systems

Koont, S. (2007). Urban agriculture in Cuba: Of, by, and for the Barrio. *Nature, Society, and Thought*, *20*(3/4), 311–325.

Koont, S. (2008). A Cuban success story: Urban agriculture. *Review of Radical Political Economics*, *40*(3), 285–291, https://doi.org/10.1177/0486613408320016

Koont, S. (2009). The urban agriculture of Havana. *Monthly Review*, *60*(1), 63–72, https://doi.org/10.14452/mr-060-08-2009-01_5

Koont, S. (2011). *Sustainable urban agriculture in Cuba*. Gainesville: The University Press of Florida, https://doi.org/10.5744/florida/9780813037578.001.0001

Kreinecker, P. (2000). La paz: Urban agriculture in harsh ecological conditions. In *Growing cities, growing food: Urban agriculture on the policy agenda. A reader on urban agriculture*, eds. N. Bakker, M. Dubbeling, S. Guendel, U. Sabel-Koschella, & H. De Zeeuw. Deutsche Stiftung fur Internationale Entwicklung (DSE), Zentralstelle fur Ernahrung und Landwirtschaft, 391–411, https://ruaf.org/document/growing-cities-growing-food/

Kuria Gathuru, M. N., Karanja, N., & Munyao, P. (2009). Gender perspectives in organic waste recycling for urban agriculture in Nairobi, Kenya. In *Women feeding cities: Mainstreaming gender in urban agriculture and food security*, eds. A. Hovorka, H. De Zeeuw, & M. Njenga. CTA/Practical Action, 141–155, https://doi.org/10.3362/9781780440460.009

Lal, R. (2020). Home gardening and urban agriculture for advancing food and nutritional security in response to the COVID-19 pandemic. *Food Security*, 1–6, https://doi.org/10.1007/s12571-020-01058-3

Lau, H. L. (2013). *Evolution of urban agriculture in Hong Kong: Stepping towards multifunctionality* (Doctoral dissertation, Chinese University of Hong Kong).

Lawson, L. J. (2005). *City bountiful: A century of community gardening in America*. University of California Press, https://doi.org/10.5860/choice.43-1557

Le, N. P., & Dung, N. M. (2018). Multifunctionality of peri-urban agriculture: A case study in Trau Quy Commune, Hanoi City. *International Journal of Rural Development*, *2*, 8–19, https://doi.org/10.22161/ijreh.2.4.2

Lefebvre, H. (1968). *Le droit à la ville*. París: An, https://doi.org/10.3406/homso.1967.1063

Leitgeb, F., Schneider, S., & Vogl, C. R. (2016). Increasing food sovereignty with urban agriculture in Cuba. *Agriculture and Human Values*, *33*(2), 415–426, https://doi.org/10.1007/s10460-015-9616-9

Lerner, A. M., & Appendini, K. (2011). Dimensions of peri-urban maize production in the Toluca-Atlacomulco Valley, Mexico. *Journal of Latin American Geography*, 87–106, https://doi.org/10.1353/lag.2011.0033

Lerner, A. M., Eakin, H., & Sweeney, S. (2013). Understanding peri-urban maize production through an examination of household livelihoods in the Toluca Metropolitan Area, Mexico. *Journal of Rural Studies*, *30*, 52–63, https://doi.org/10.1016/j.jrurstud.2012.11.001

Lerner, A., Sweeney, S., & Eakin, H. (2014). Growing buildings in corn fields: Urban expansion and the persistence of maize in the Toluca Metropolitan Area, Mexico. *Urban Studies*, *51*(10), 2185–2201, https://doi.org/10.1177/0042098013506064

Lima, P. T., Sanchez, L. M. R., & García, B. I. (2000). Mexico City: The integration of urban agriculture to contain urban sprawl. In *Growing cities, growing food: Urban agriculture on the policy agenda. A reader on urban agriculture*, eds. N. Bakker, M. Dubbeling, S. Guendel, U. Sabel-Koschella, & H. De Zeeuw. Deutsche Stiftung fur Internationale Entwicklung (DSE), Zentralstelle fur Ernahrung und Landwirtschaft, 363–390, https://ruaf.org/document/growing-cities-growing-food/

Liu, T., Yang, M., Han, Z., & Ow, D. W. (2016). Rooftop production of leafy vegetables can be profitable and less contaminated than farm-grown vegetables. *Agronomy for Sustainable Development*, *36*(3), 1–9, https://doi.org/10.1007/s13593-016-0378-6

Lohrberg, F., Lička, L., Scazzosi, L., & Timpe, A. (2016). *Urban agriculture Europe*. Berlin: Jovis.

Losada, H., Bennett, R., Soriano, R., Vieyra, J., & Cortes, J. (2000). Urban agriculture in Mexico City: Functions provided by the use of space for dairy based livelihoods. *Cities*, *17*(6), 419–431, https://doi.org/10.1016/S0264-2751(00)00041-X

Losada, H., Martínez, H., Vieyra, J., Pealing, R., Zavala, R., & Cortés, J. (1998). Urban agriculture in the metropolitan zone of Mexico City: Changes over time in urban, sub-urban and peri-urban areas. *Environment and Urbanization*, *10*(2), 37–54, https://doi.org/10.4324/9781315800486-13

Lovell, S. T. (2010). Multifunctional urban agriculture for sustainable land use planning in the United States. *Sustainability*, *2*(8), 2499–2522, https://doi.org/10.1201/b18713-20

Lydecker, M., & Drechsel, P. (2010). Urban agriculture and sanitation services in Accra, Ghana: the overlooked contribution. *International Journal of Agricultural Sustainability*, *8*(1–2), 94–103, https://doi.org/10.3763/ijas.2009.0453

Lynch, K., Maconachie, R., Binns, T., Tengbe, P., & Bangura, K. (2013). Meeting the urban challenge? Urban agriculture and food security in post-conflict Freetown, Sierra Leone. *Applied Geography*, *36*, 31–39, https://doi.org/10.1016/j.apgeog.2012.06.007

Madaleno, I. (2000). Urban agriculture in Belém, Brazil. *Cities*, *17*(1), 73–77, https://doi.org/10.1016/s0264-2751(99)00053-0

Maldonado Villavicencio, L. (2009). Urban agriculture as a livelihood strategy in Lima, Peru. In *Agriculture in urban planning generating livelihoods and food security*, ed. M. Redwood. London: IRDC, 49–70, https://doi.org/10.4324/9781849770439

Marshall, F., & Randhawa, P. (2017). *India's peri-urban frontier: Rural-urban transformations and food security.* International Institute of Environment and Development (IIED).

Maxwell, D. G. (1995). Alternative food security strategy: A household analysis of urban agriculture in Kampala. *World Development*, *23*(10), 1669–1681, https://doi.org/10.1016/0305-750x(95)00073-l

Maxwell, D., & Armar-Klemesu, M. (1998). *Urban agriculture: introduction and review of literature.* Accra: Noguchi Memorial Institute for Medical Research.

Maxwell, D., Levin, C., & Csete, J. (1998). Does urban agriculture help prevent malnutrition? Evidence from Kampala. *Food Policy*, *23*(5), 411–424, https://doi.org/10.1016/s0306-9192(98)00047-5

Mboganie-Mwangi, A., & Foeken, D. W. J. (1996). Urban agriculture, food security and nutrition in low income areas of the city of Nairobi, Kenya. *African Urban Quarterly*, *11*(2/3), 170–179.

McClintock, N., Miewald, C., & McCann, E. (2021). Governing urban agriculture: Formalization, resistance and re-visioning in two 'Green' cities. *International Journal of Urban and Regional Research*, *45*(3), 498–518, https://doi.org/10.1111/1468-2427.12993

Midmore, D. J., & Jansen, H. G. (2003). Supplying vegetables to Asian cities: Is there a case for peri-urban production? *Food Policy*, *28*(1), 13–27, https://doi.org/10.1016/s0306-9192(02)00067-2

Mkwambisi, D. D., Fraser, E. D., & Dougill, A. J. (2011). Urban agriculture and poverty reduction: Evaluating how food production in cities contributes to food security, employment and income in Malawi. *Journal of International Development*, *23*(2), 181–203, https://doi.org/10.1002/jid.1657

Mok, H. F., Williamson, V. G., Grove, J. R., Burry, K., Barker, S. F., & Hamilton, A. J. (2014). Strawberry fields forever? Urban agriculture in developed countries: A review. *Agronomy for Sustainable Development*, *34*(1), 21–43, https://doi.org/10.1007/s13593-013-0156-7

Mougeot, L. J. (2000). Urban agriculture: Definition, presence, potentials and risks, and policy challenges. *Cities feeding people series*. IDRC.

Mrozik, W., Vinitnantharat, S., Thongsamer, T., Pansuk, N., Pattanachan, P., Thayanukul, P., & Werner, D. (2019). The food-water quality nexus in periurban aquacultures downstream of Bangkok, Thailand. *Science of the Total Environment*, *695*, 1–10, https://doi.org/10.1016/j.scitotenv.2019.133923

Murphy, C. (1999). *Cultivating Havana: Urban agriculture and food security in the years of crisis.* Food First Institute for Food and Development Policy.

Mutonodzo, C. (2009). The social and economic implications of urban agriculture on food security in Harare, Zimbabwe. *Agriculture in Urban Planning Generating Livelihoods and Food Security*, ed. M. Redwood. London: IRDC, 73–89, https://doi.org/10.4324/9781849770439

Nabulo, G., Black, C. R., Craigon, J., & Young, S. D. (2012). Does consumption of leafy vegetables grown in peri-urban agriculture pose a risk to human health? *Environmental Pollution*, *162*, 389–398, https://doi.org/10.1016/j.envpol.2011.11.040

Nabulo, G., Kiguli, J., & Kiguli, L. (2009). Gender in urban crop production in hazardous areas in Kampala, Uganda. In *Women feeding cities: Mainstreaming gender in urban agriculture and food security*, eds. A. Hovorka, H. De Zeeuw, & M. Njenga. CTA/Practical Action, 79–92, https://doi.org/10.3362/9781780440460.004

Nambi, A. A., Rengalakshmi, R., Madhavan, M., & Venkatachalam, L. (2014). *Building urban resilience: Assessing urban and peri-urban agriculture in Chennai, India.* United Nations Environment Programme (UNEP), Nairobi, Kenya, 11–23.

Nelson, E., Scott, S., Cukier, J., & Galán, Á. L. (2009). Institutionalizing agroecology: Successes and challenges in Cuba. *Agriculture and Human Values*, *26*(3), 233–243, https://doi.org/10.1007/s10460-008-9156-7

Nettle, C. (2016). *Community gardening as social action*. London: Routledge, https://doi.org/10.4324/9781315572970

Nicholls, E., Ely, A., Birkin, L., Basu, P., & Goulson, D. (2020). The contribution of small-scale food production in urban areas to the sustainable development goals: A review and case study. *Sustainability Science*, 1–15, https://doi.org/10.1007/s11625-020-00792-z

Njogu, E. W. (2012). Household food security among urban farmers in Nairobi, Kenya. In *Agriculture in urban planning generating livelihoods and food security*, ed. M. Redwood. London: IRDC, 21–34, https://doi.org/10.4324/9781849770439

Novo, M. G., & Murphy, C. (2000). Urban agriculture in the city of Havana: A popular response to a crisis. In *Growing cities, growing food: Urban agriculture on the policy agenda. A reader on urban agriculture*, eds. N. Bakker, M. Dubbeling, S. Guendel, U. Sabel-Koschella, & H. De Zeeuw. Deutsche Stiftung fur Internationale Entwicklung (DSE), Zentralstelle fur Ernahrung und Landwirtschaft, 329–346, https://ruaf.org/document/growing-cities-growing-food/

Opitz, I., Berges, R., Piorr, A., & Krikser, T. (2016). Contributing to food security in urban areas: differences between urban agriculture and peri-urban agriculture in the Global North. *Agriculture and Human Values*, *33*(2), 341–358, https://doi.org/10.1007/s10460-015-9610-2

Pearson, L. J., Pearson, L., & Pearson, C. J. (2010). Sustainable urban agriculture: Stocktake and opportunities. *International Journal of Agricultural Sustainability*, *8*(1–2), 7–19, https://doi.org/10.3763/ijas.2009.0468

Pham, V. C., Pham, T. T. H., Tong, T. H. A., Nguyen, T. T. H., & Pham, N. H. (2015). The conversion of agricultural land in the peri-urban areas of Hanoi (Vietnam): Patterns in space and time. *Journal of Land Use Science*, *10*(2), 224–242, https://doi.org/10.1080/1747423x.2014.884643

Piacentini, R. D. N., Bracalenti, L., Salum, G. M., Zimmermann, E. D., Terriles, R. A., Bartolomé, S., & Coronel, A. (2014). Monitoring the climate change impacts of urban agriculture in Rosario, Argentina. *Urban Agriculture Magazine*, *27*, 50–53.

Plonska, O. (2017). *This garden is who I am. How urban gardeners in Cuba experience their relationship with the state, their gardens and moments of freedom*. VU University of Amsterdam.

Ponce, M., & Donoso, L. (2009). Urban agriculture as a strategy to promote equality of opportunities and rights for men and women in Rosario, Argentina. In *Women feeding cities: Mainstreaming gender in urban agriculture and food security*, eds. A. Hovorka, H. De Zeeuw, & M. Njenga. CTA/Practical Action, 157–179, https://doi.org/10.3362/9781780440460.010

Poulsen, M. N., Neff, R. A., & Winch, P. J. (2017). The multifunctionality of urban farming: Perceived benefits for neighbourhood improvement. *Local Environment*, *22*(11), 1411–1427, https://doi.org/10.1080/13549839.2017.1357686

Prain, G., Karanja, N., & Lee-Smith, D. (2010). *African urban harvest: Agriculture in the cities of Cameroon, Kenya and Uganda*. IDRC, Ottawa, ON, CA, https://doi.org/10.1007/978-1-4419-6250-8

Premat, A. (2003). Small-scale urban agriculture in Havana and the reproduction of the "new man" in contemporary Cuba. *European Review of Latin American and Caribbean Studies*, 85–99, https://doi.org/10.18352/erlacs.9695

Premat, A. (2005). Moving between the plan and the ground: Shifting perspectives on urban agriculture in Havana, Cuba. In *Agropolis*, ed. L. J. A. Mougeot, 171–204, https://idrc-crdi. ca/en/book/agropolis-social-political-and-environmental-dimensions-urban-agriculture

Premat, A. (2009). State power, private plots and the greening of Havana's urban agriculture movement. *City & Society*, 21(1), 28–57, https://doi.org/10.1111/j.1548-744x.2009.01014.x

Premat, A. (2012). *Sowing change: the making of Havana's urban agriculture.* Vanderbilt University Press, https://doi.org/10.2307/j.ctv16759bh.7

Propersi, P. S. (2008). The health impacts of farming on producers in Rosario, Argentina. In *Agriculture in urban planning generating livelihoods and food security*, ed. M. Redwood. London: IRDC, 185–198, https://doi.org/10.4324/9781849770439

Pulliat, G. (2015). Food securitization and urban agriculture in Hanoi (Vietnam). *Articulo-Journal of Urban Research*, 7, https://doi.org/10.4000/articulo.2845

Ratta, A., & Nasr, J. (1996). Urban agriculture and the African urban food supply system. *African Urban Quarterly*, 11(2/3), 154–161.

Redwood, M. (2012). *Agriculture in urban planning generating livelihoods and food security.* London: IRDC, https://doi.org/10.4324/9781849770439

Rodríguez Castellón, S. (2003). La agricultura urbana y la producción de alimentos: La experiencia de Cuba [Urban agriculture and food production: The Cuban experience]. *Cuba Siglo XXI, 30,* 77–101.

Safayet, M., Arefin, M. F., & Hasan, M. M. U. (2017). Present practice and future prospect of rooftop farming in Dhaka city: A step towards urban sustainability. *Journal of Urban Management, 6*(2), 56–65, https://doi.org/10.1016/j.jum.2017.12.001

Sánchez, H. (2009). Periurbanización y espacios rurales en la periferia de las ciudades. *Procuraduría Agraria, Estudios Agrarios, 15*(4), 93–123.

Scoones, I. (1998). *Sustainable rural livelihoods: A framework for analysis.* IDS Working Paper 72, 1–22.

Shillington, L. J. (2013). Right to food, right to the city: Household urban agriculture, and socionatural metabolism in Managua, Nicaragua. *Geoforum, 44*, 103–111, https://doi. org/10.1016/j.geoforum.2012.02.006

Simón Reardon, J. A., & Pérez, R. A. (2010). Agroecology and the development of indicators of food sovereignty in Cuban food systems. *Journal of Sustainable Agriculture, 34*(8), 907–922, https://doi.org/10.1080/10440046.2010.519205

Smit, J., & Nasr, J. (1992). Urban agriculture for sustainable cities: Using wastes and idle land and water bodies as resources. *Environment and Urbanization, 4*(2), 141–152, https://doi.org/10.1177/095624789200400214

Smit, J., Nasr, J., & Ratta, A. (2001). *Urban agriculture: Food, jobs and sustainable cities.* New York, http://www.jacsmit.com/book.html

Soto, N., Merzthal, G., Ordoñez, M., & Touzet, M. (2006). Urban agriculture, poverty alleviation, and gender in Villa María del Triunfo, Peru. In *Women feeding cities: Mainstreaming gender in urban agriculture and food security*, eds. A. Hovorka, H. De Zeeuw, & M. Njenga. CTA/Practical Action, 124–139, https://doi.org/10.3362/9781780440460.008

Specht, K., Siebert, R., Hartmann, I., Freisinger, U. B., Sawicka, M., Werner, A., & Dierich, A. (2014). Urban agriculture of the future: an overview of sustainability aspects of food production in and on buildings. *Agriculture and Human Values, 31*(1), 33–51, https://doi. org/10.1007/s10460-013-9448-4

Specht, K., Siebert, R., & Thomaier, S. (2016). Perception and acceptance of agricultural production in and on urban buildings (ZFarming): A qualitative study from Berlin, Germany. *Agriculture and Human Values, 33*(4), 753–769, https://doi.org/10.1007/s10460-015-9658-z

Speybroeck, N., Berkvens, D., Mfoukou-Ntsakala, A., Aerts, M., Hens, N., Van Huylenbroeck, G., & Thys, E. (2004). Classification trees versus multinomial models in the analysis of urban farming systems in Central Africa. *Agricultural Systems, 80*(2), 133–149, https://doi.org/10.1016/j.agsy.2003.06.006

Spiaggi, E. (2010). Urban agriculture and local sustainable development in Rosario, Argentina: Integration of economic, social, technical and environmental variables. In *Agropolis*, ed. L. J. A. Mougeot. London: Routledge, 205–220, https://idrc-crdi.ca/en/book/agropolis-social-political-and-environmental-dimensions-urban-agriculture

Srikanth, R., & Naik, D. (2004). Prevalence of Giardiasis due to wastewater reuse for agriculture in the suburbs of Asmara City, Eritrea. *International Journal of Environmental Health Research, 14*(1), 43–52, https://doi.org/10.1080/0960312031000163391 2

Stoler, J., Weeks, J. R., Getis, A., & Hill, A. G. (2009). Distance threshold for the effect of urban agriculture on elevated self-reported malaria prevalence in Accra, Ghana. *The American Journal of Tropical Medicine and Hygiene, 80*(4), 547–554, https://doi.org/10.4269/ajtmh.2009.80.547

Su, Y. L., Wang, Y. F., & Ow, D. W. (2020). Increasing effectiveness of urban rooftop farming through reflector-assisted double-layer hydroponic production. *Urban Forestry & Urban Greening, 54*, 1–9, https://doi.org/10.1016/j.ufug.2020.126766

Tano, B. F., Abo, K., Dembele, A., & Fondio, L. (2011). Systèmes de production et pratiques à risque en agriculture urbaine: cas du maraîchage dans la ville de Yamoussoukro en Côte d'Ivoire [Production systems and risky practices in urban agriculture: Case of market gardening in the city of Yamoussoukro in Côte d'Ivoire]. *International Journal of Biological and Chemical Sciences, 5*(6), 2317–2329, https://doi.org/10.4314/ijbcs.v5i6.12

Thapa, R. B. (2003). *Spatial decision support model for sustainable peri-urban agriculture: Case study of Hanoi Province, Vietnam* (Doctoral dissertation, MSc. Thesis. Asian Institute of Technology).

Thapa, R. B., Borne, F., Kusanagi, M., & Pham, V. C. (2004). Integration of RS, GIS and AHP for Hanoi peri-urban agriculture planning. *Proceedings of the Map Asia*. Beijing.

Thapa, R. B., & Murayama, Y. (2008). Land evaluation for peri-urban agriculture using analytical hierarchical process and geographic information system techniques: A case study of Hanoi. *Land Use Policy, 25*(2), 225–239, https://doi.org/10.1016/j.landusepol.2007.06.004

Thomaier, S., Specht, K., Henckel, D., Dierich, A., Siebert, R., Freisinger, U. B., & Sawicka, M. (2015). Farming in and on urban buildings: Present practice and specific novelties of Zero-Acreage Farming (ZFarming). *Renewable Agriculture and Food Systems, 30*(1), 43–54, https://doi.org/10.1017/s1742170514000143

Toriro, P. (2009). Gender dynamics in the Musikavanhu urban agriculture movement, Harare, Zimbabwe. *Women feeding cities: Mainstreaming gender in urban agriculture and food security*, eds. A. Hovorka, H. De Zeeuw, & M. Njenga. CTA/Practical Action, 93–104, https://doi.org/10.3362/9781780440460.006

Trembecka, A., & Kwartnik-Pruc, A. (2018). An analysis of the changes in the structure of allotment gardens in Poland and of the process of regulating legal status. *Sustainability, 10*(11), 3829, https://doi.org/10.3390/su10113829

Valley, W., & Wittman, H. (2019). Beyond feeding the city: The multifunctionality of urban farming in Vancouver, BC. *City, Culture and Society, 16*, 36–44, https://doi.org/10.1016/j.ccs.2018.03.004

Van Rooijen, D. J., Biggs, T. W., Smout, I., & Drechsel, P. (2010). Urban growth, wastewater production and use in irrigated agriculture: A comparative study of Accra, Addis

Ababa and Hyderabad. *Irrigation and Drainage Systems*, *24*(1-2), 53–64, https://doi.org/10.1007/s10795-009-9089-3

Van Veenhuizen, R. (2006). *Cities farming for the future: Urban agriculture for green and productive cities*. Ottawa, Canada: RUAF Foundation, IIRR, IDRC.

Viljoen, A., & Bohn, K. (2014). *Second nature urban agriculture: Designing productive cities*. London: Routledge, https://doi.org/10.4324/9781315771144

Viljoen, A., & Howe, J. (2005). *Continuous productive urban landscapes*. London: Routledge, https://doi.org/10.4324/9780080454528

Wang, T., & Pryor, M. (2019). Social value of urban rooftop farming: A Hong Kong case study. *Agricultural Economics-Current Issues*, https://doi.org/10.5772/intechopen.89279

White, M. M. (2011). Sisters of the soil: Urban gardening as resistance in Detroit. *Race/Ethnicity: Multidisciplinary Global Contexts*, *5*(1), 13–28, https://doi.org/10.2979/racethmulglocon.5.1.13

Yang, Z., Cai, J., & Sliuzas, R. (2010). Agro-tourism enterprises as a form of multi-functional urban agriculture for peri-urban development in China. *Habitat International*, *34*(4), 374–385, https://doi.org/10.1016/j.habitatint.2009.11.002

Yeung, Y. M. (1987). Examples of urban agriculture in Asia. *Food and Nutrition Bulletin*, *9*(2), 1–10.

Zasada, I. (2011). Multifunctional peri-urban agriculture – a review of societal demands and the provision of goods and services by farming. *Land Use Policy*, *28*(4), 639–648, https://doi.org/10.1016/j.landusepol.2011.01.008

Zasada, I., Schmutz, U., Wascher, D., Kneafsey, M., Corsi, S., Mazzocchi, C., & Piorr, A. (2019). Food beyond the city – analysing foodsheds and self-sufficiency for different food system scenarios in European metropolitan regions. *City, Culture and Society*, *16*, 25–35, https://doi.org/10.1016/j.ccs.2017.06.002

Zezza, A., & Tasciotti, L. (2010). Urban agriculture, poverty, and food security: Empirical evidence from a sample of developing countries. *Food Policy*, *35*(4), 265–273, https://doi.org/10.1016/j.foodpol.2010.04.007

Zhong, C., Hu, R., Wang, M., Xue, W., & He, L. (2020). The impact of urbanization on urban agriculture: Evidence from China. *Journal of Cleaner Production*, *276*, 122686, https://doi.org/10.1016/j.jclepro.2020.122686

3 Havana – a case study of urban agriculture in Latin America

Havana, the capital city of Cuba and one of the largest metropolises in the Caribbean, is an example of a city where dynamic development of urban agriculture occurred relatively recently – in the last decade of the 20th century. Due to the large number and ubiquity of urban gardens in the city space as well as the crucial role they play in the urban ecosystem, Havana should be treated as one of the most interesting examples of urban agriculture in all of Latin America. Choosing Cuba's capital city as the subject of research on urban agriculture enables an in-depth analysis of the history of development of an alternative intra-urban food system from the time of its establishment within the metropolis (Chaplowe 1998; Premat 2005; Giradet 2012).

The detailed analysis of Cuba's case, taken up in this book, covered an arbitrary area of Havana's densely built-up space of 76.7 km². A total of 55 urban gardens were classified on the basis of satellite and aerial images of the area.[1] Fifty of them were visited during the field research carried out in May 2018. The five remaining sites were located in a fenced-off and closed-off military area or inaccessible private areas, so their analysis was impossible. During field verification, 43 out of 50 sites were classified as allocated for urban agriculture (with visible crops). These sites underwent further detailed analysis. Among the remaining seven sites, five were identified as abandoned between the time when the satellite image was captured and the time of field research. The two remaining sites had been erroneously classified at the stage of satellite image analysis. This error accounts for 4% of all visited sites, which is below the adopted permissible error margin of 5%.

During the field research, 21 semi-structured interviews were conducted. The group of respondents comprised employees of the individual gardens, that is, their managers, gardeners and sales point employees. Detailed questions asked during the interviews concerned the functioning of the gardens. The answers provided a lot of valuable information which would have been impossible to obtain on the basis of observations alone, including information on the number of employees, a complete list of plants grown or compost production sites. Since it was possible for both the respondent and the interviewer to introduce additional content (the selected research method of a semi-structured interview allows for it), the scope of information collected considerably exceeded the previously prepared list of issues. This, in turn, enabled comprehensive characterisation of the analysed gardens. Due to the conditions

DOI: 10.4324/9781003429845-3

of carrying out research in Cuba, these interviews were not recorded. After the end of each interview, the respondents' answers were entered into previously prepared electronic questionnaire forms.[2] Information obtained during the interviews with actors involved in urban agriculture and the accompanying field observations allowed for the characteristics of each of the analysed gardens to be determined.

A detailed presentation of the results of the field research carried out will be preceded by a discussion of issues depicting various contexts of the analysed subject matter. First, the features of the spatial and functional structure of Havana will be presented, with particular emphasis on its spatial development and social and economic transformation during the period since the city's founding in 1519 until the present day. This will be followed by a presentation of the history of urban agriculture in Cuba as well as factors which led to its transformation from a grassroots movement of residents to a highly institutionalised activity. The last sub-chapter preceding detailed results of the field research is a critical analysis of typology of urban agriculture in Cuba used in literature.

3.1   Spatial and functional structure of the city

Havana (Sp. *La Habana*) is situated on the Havana Bay, on the northern coast of Cuba. The administrative area of the city covers 727 km², which amounts to 0.6% of the total area of the country (Colantonio & Potter 2006). According to the Köppen climate classification, Havana has a tropical climate. Due to its location on an island and presence of breeze, the conditions are less extreme than in inland areas in the same climate zone. Nevertheless, similar to other islands with marine-island tropical climate, the summer period in Cuba is characterised by high air temperatures and high relative humidity. In August, the average maximum daily temperature is 31.4 C, while the average maximum daily relative humidity amounts to 91% (Tablada et al. 2009). The city is characterised by a dense river system. The largest river running through Havana, from south to north, is the Almendares, with forests and marshes concentrated along its banks. The terrain height differences do not exceed 100 m and there are several small elevations within the city. In 2023, the Cuban capital was inhabited by 2.15 million people – approximately 19% of the country's population.[3] Until the 1960s, Havana was an example of a primate city since the number of its residents was 3.5 times higher than the total population of two other largest Cuban cities – Santiago de Cuba and Camagüey (Colantonio & Potter 2006).

When analysing the spatial and functional structure of the Cuban capital city, one should take a look at the factors which have shaped its current characteristics. This requires discussing the history of the city's development. It can be divided into four basic periods: colonial city (1514–1898), city in times of a pseudo-republic (1898–1959), city after the revolution (1959–1989) and the "Special Period" (from 1990 until the present day) (Colantonio & Potter 2006).

3.1.1   *Colonial city (1514–1898)*

Havana was founded in the north of Cuba by the Spanish already in 1519, after the settlement of San Cristóbal de la Habana (which dates back to 1514) was moved from the south to the north coast of the island (Colantonio & Potter 2006; Currie 2012). At the time of "colonial meeting", the small population of local indigenous people was exterminated. No traces of pre-Columbian architecture have survived to this day either (Coyula Hamberg 2003). Havana was constructed according to the Laws of the Indies (*Las Leyes de Indias*) – a set of city planning rules in Spanish colonies in both Americas and the Philippines (Rojas 1977). The document (its full names is *Ordenanzas de Descubrimiento, Nueva Población y Pacificación de las Indias* – literally: Regulations on Discovery, New Population and Pacification of the Indies) was adopted in 1573 by Philip II of the Habsburg dynasty (king of Spain in 1556–1598). It included and organised a number of previously existing and unrelated laws, specifying the mechanisms of colonial city planning, including 148 regulations on spatial planning itself (Rojas 1977; Del Vas Mingo 1985; Sánchez Bella 1992; Melendo & Verdejo 2008). According to the Laws of the Indies, a necessary condition when founding any new settlement was its organisation around a central point, that is, a rectangular city square, from which main streets extended in an orthogonal layout. Moreover, its surface area was supposed to be adjusted to the assumed population of the city. The document also determined the location of the church, the residence of the authorities and the administration, as well as commercial facilities around the square. What is more, every settlement was supposed to be "organic" and adapted to the local climate conditions. For instance, cities located in the subtropical or tropical climate zone should have narrow streets that provide shade (Melendo & Verdejo 2008). Introduction of the Laws of the Indies was a key factor determining the morphology of Spanish colonial cities, which to this day exhibit numerous similarities in the urban layout of their historical districts. Central squares and the orthogonal grid of parallel and perpendicular streets extending from them, setting out quarters of densely packed buildings, can be found not only in Cuba, but also in Caracas, Venezuela, in Lima, Peru, in Buenos Aires, Argentina, in Mexico or Panama.

Thanks to its location on the silver trail, leading from South America through Panama and the Caribbean to Europe, Havana started playing a key role in transport of goods between the Old and New World already in the first years of its existence. As the city increased in prominence, the Spanish Crown ordered construction of a number of fortifications intended to protect the coast. The most important of those forts were *La Fuerza* (built in 1558–1577), *Tres Reyes del Morro* (1589–1610) and *San Salvador de la Punta* (1589–1600) (Colantonio & Potter 2006). In the early 17th century, Havana had approximately 4,000 residents and was the most important outpost for many Spanish sailors and soldiers (Coyula Hamberg 2003). By the end of the 17th century, the city had become one of the most prominent transit ports in all of Latin America (Coyula Hamberg 2003; Currie 2012). For centuries, it was the reloading site for luxurious goods from the entire region, such as silver from Bolivia or gold from Columbia (known as *Castilla de oro* – literally, the Castile of gold – at the time), which was subsequently exported to Spain (Wolf 1982). At the end of the 18th century, thanks to the development of plantations based on the work of slaves brought from Africa, Cuba became one of the main suppliers of sugarcane to the European market (ibidem). Havana, in turn, profited the most from sugar production, which contributed to its dynamic development (Segre & Baroni 1998).

The island's growing commercial position was also reflected in the city's architecture (Coyula Hamberg 2003). Numerous representative public buildings, private residences of "sugar barons" and recreational public spaces appeared within the city during that time (Coyula Hamberg 2003; Currie 2012).

An important urbanistic change at the end of the 18th century was transformation of the vicinity of *Plaza de Armas* (in the *Habana Vieja* district), which had previously been a military area, into a public space full of greenery. During the same period, two wide, representative promenades were also set out: *Alameda de Paula* and *Alameda de Extramuros* (currently *Paseo del Prado*). Both projects made Havana more reminiscent of European cities. The spaces created thanks to that, especially transformation of military zones into public ones, had a significant impact on the evolution of Havana's spatial and functional structure (Hernández 2011).

In the early 19th century, development of Spanish Havana accelerated in connection with the breakup of Spain's colonial empire in the New World, where Cuba became one of the last two Spanish colonies in America (Coyula Hamberg 2003).[4] From that point on, key initiatives undertaken by the government in Madrid in relation to both Americas were focused around Havana. Moreover, during that time, the city gained many new residents – primarily liberated slaves. This led to a dynamic spatial expansion of the city and its sprawl into the nearby rural areas. However, it should be noted that this development was radial, occurring mainly along the newly created thoroughfares (Sp. *calzadas*) – wide, long streets leading out of the colonial centre. They were set out towards the south and south-east (Currie 2012). According to the urbanist assumption called *Plan de Ensanche* (translation: Expansion Plan), implemented in 1817–1819 under the supervision of Colonel-Engineer Antonio María de la Torre, streets were supposed to connect the city, in particular its colonial walls, to the rural hinterland in an orderly manner (Hernández 2011). Areas along the thoroughfares were subsequently gradually developed (Coyula Hamberg 2003). In the first half of the 19th century, the land adjacent to *Calzada del Cerro* was occupied by representatives of the so-called sugar aristocracy and baroque villas were largely replaced by detached neoclassicist villas with vast gardens and porches (Hernández 2011). In the second half of the 19th century (in 1863), the walls which surrounded the city and restricted its development were demolished. The modern district of *El Vedado*, located closer to the coast and designed by Luis Yboleón in 1859, started playing an increasingly significant role. It featured a lot of greenery and distinctive trees planted along the roads making up a regular street grid. It was *El Vedado* that became the new destination for the increasingly numerous local elites, which contributed to the district's dynamic development after Cuba gained independence (Coyula Hamberg 2003).

3.1.2   The city in times of pseudo-republic (1898–1959)

In 1898, the Spanish-American War broke out, resulting in the island being taken over by the United States. Cuba remained an American protectorate until 1902. Therefore, it obtained formal independence from the United States, not Spain. The period between Cuba gaining independence in 1902 and the Cuban Revolution, which ended with Communist victory in 1959, is referred to in literature as the pseudo-republican period due to the strong political, economic and cultural influence of Americans (Colantonio & Potter 2006). During this period, Havana underwent intensive spatial development and an uncontrolled construction boom (Coyula Hamberg 2003; Ponce Herrero 2007b), resulting from the influx of American investment capital. The chaotic development of Havana in the first decades of the 20th century was first and foremost due to the fact that in 1902, it became the capital city of the newly established Republic of Cuba. Second, the lack of adequate planning documentation and legal regulations (which was typical of young countries) hampered supervision over the city's spatial development by the authorities. This allowed for land to be purchased by foreign developers, which ultimately contributed to the city's unchecked expansion (Colantonio & Potter 2006; Hernández 2011). Another reason for spatial expansion was common adoption of the American development model based on car transport (Ponce Herrero 2007b; Hernández 2011). What is more, the early 20th century was also a period of intensive industrialisation of Cuba, which mainly consisted in development of the textile, food and construction industries, which also had a significant impact on the city's morphology.

The industrialisation-driven development of the city in the 20th century took the following directions: south-west (Rancho Boyeros expansion axis), south-east (Cotorro expansion axis) and east (Regla expansion axis) (Colantonio & Potter 2006). These areas, characterised by mixed forms of land use, comprised mainly of local industrial plants and workshops surrounded by workers' apartments. During the period preceding the Cuban Revolution, Havana had at least four fully developed functional centres (or hubs). The first one was the colonial *Habana Vieja*, whose form has remained practically unchanged since the Spanish times (ibidem). It is the densest part of the city, built on a chequerboard plan and characterised by a regular grid of narrow streets intersecting at a right angle as well as numerous squares, next to which churches were built. Since its establishment, *Habana Vieja* has invariably fulfilled a political (administrative), residential, cultural and commercial functions. The second centre is the east fragment of today's *Centro Habana* district, developed in the early 20th century and called the First Republican Centre. It fulfilled an administrative, commercial as well as a recreational function due to the presence of facilities such as the Grand Theatre of Havana (Sp. *Gran Teatro de La Habana*), which opened in 1915. The area considered to be the Second Republican Centre is the vicinity of the former Civic Square (Sp. *Plaza Cívica*), comprising mainly governmental buildings. It was renamed the Revolution Square (Sp. *Plaza de la Revolución*) after 1959 (Alonso González 2016). Its construction commenced during Fulgencio Batista's dictatorship (1933–1959). The two other complementary recreational and tourist hubs, which date back to the second half of the 19th century, are *El Vedado* and *Miramar*. The latter is characterised by larger quarters and more greenery (Coyula Hamberg 2003; Colantonio & Potter 2006). The *Miramar* district, which was intended for the richest residents of Havana, was left purposefully free from commercial facilities to limit the migration of poorer population (Coyula Hamberg 2003; Currie 2012).

Apart from the three industrial sectors, during the period of the pseudo-republic, the city also expanded along the east and west coast of the Straits of Florida. This was mainly the result of development of luxurious tourist infrastructure for Cuban and American elites (Colantonio & Potter 2006). An important feature of the city's development during the pseudo-republican period is also the clear spatial functional and class segregation (Ponce Herrero 2007b).

The lowest social class, called subproletariat by G. Ponce Herrero (2007b), was concentrated in the oldest colonial district of *Habana Vieja*, where the main financial centre was also located, as well as in the port and industrial areas along the west and south-west coast of the Havana Bay. To the south, in turn, was the area occupied by the proletariat. The working class also inhabited the district of *La Regla* on the other side of the bay. The elites converged mainly on the coast, which predominantly fulfilled the tourist function. However, it should be emphasised that it was the *Miramar* district where the upper class resided. The east part of the coast was occupied by the middle class because the East Beaches (Sp. *Playas del Este*) had lost their prestigious significance at the time. The middle class, which separated the proletariat from the elites, also resided in areas stretching towards the south from the First Republican Centre, through the southern part of the *El Vedado* district all the way to the Second Centre. Spatial segregation noted by G. Ponce Herrero (2007b) was so strongly entrenched in the urban tissue that it is visible to this day.

To manage the dynamic, uncontrolled development of the city and prevent degradation of the existing districts, in 1954, Havana planners led by the city's chief architect Martínez Inclán introduced the so-called Havana Charter (Sp. *La Carta de la Habana*). It was based on the vision of a modernist city, published in 1933 and included in the Athens Charter by Le Corbusier (Le Corbusier 1933). Utilising the ideas of one of the main representatives of international modernism was a popular practice in many Latin American cities in the 1950s, for example, Buenos Aires, Bogotá or Brasília (Ponce Herrero 2007b). The plans included, among others, rebuilding the colonial district of *Habana Vieja*, including widening of the streets and demolishing some of the historical buildings to allow for the Capitol building[5] to be connected to the coast, building housing estates for port workers or introducing linear parks and shared pedestrian and motor zones (ibidem). Another priority was the development of the tourist sector on the coast and in the colonial parts of the city. The proposed solutions were not implemented due to the outbreak of the Cuban Revolution in 1953. Later on, modernist ideas that were popular in countries of the Eastern bloc were also introduced in Cuba, which is nowadays clearly visible in Havana in the form of distinct Soviet influences (Edge et al. 2006). Projects whose implementation commenced before the end of the Revolution were a modernist residential compound in the *Habana del Este* district and construction of the Civic Square (Sp. *Plaza Cívica*) (Ponce Herrero 2007b).

3.1.3   *The city after the revolution (1959–1989)*

Following the Cuban Revolution, carried out by the 26th of July Movement under the leadership of Fidel Castro in 1953/56–1959, the spatial planning policy

in Cuban cities changed drastically. The newly established Communist government decided to reduce the economic disproportions between Cuban cities and
rural areas, which had been a serious impediment for the country's economic
and cultural development (Colantonio & Potter 2006). Therefore, priority was
assigned to two measures aimed at transforming the spatial structure of the entire
island – ruralisation of cities (1) combined with simultaneous urbanisation of the
countryside (2) (Potter & Dann 2000; Edge et al. 2006; Ponce Herrero 2007b).
Both measures primarily amounted to investing in rural areas at the cost of
slowing down the capital city's development (Ponce Herrero 2007b; Hernández
2011). Restricting financial resources contributed to the deceleration of Havana's
spatial expansion and almost completely halted modernisation of the existing infrastructure (Colantonio & Potter 2006). Due to the previous pro-American elites
escaping the city (and the country), many villas in *El Vedado* and *Miramar* were
abandoned or, in later years, occupied by diplomatic missions (Currie 2012). Although Havana did not experience intensive infrastructure development after the
revolution, its landscape underwent some significant transformations. First, there
was an increase in the share of public spaces, planned as locations for national
celebrations and demonstrations (Edge et al. 2006). The vicinity of the Revolution Square (Sp. *Plaza de la Revolución*) (previously Civic Square) is a prominent example of this occurrence – state administration buildings surrounded by
vast swathes of greenery were constructed around that area.

Moreover, housing estates with modernist apartment buildings modelled on
Eastern European cities appeared in Havana. They were located in the outskirts
of the city or in previously undeveloped wedges separating the sectors of densely
packed industrial buildings developing along the main thoroughfares. Socialist architecture (i.e. housing estates and monumental administration buildings) as well
as the distinct, extensive urban layout (entire districts of sparsely located residential buildings) stood in striking contrast to the previous dense, baroque and neoclassicist architecture in the city centre. The lack of funds for new investments in
the historical districts of Cuban cities resulted in one outcome which was positive
from today's perspective. Indirectly, relocation of large-scale investments led to
preservation of valuable colonial architecture, which is nowadays one of the main
attractions for Cuba's nascent mass tourism (Coyula Hamberg 2003; Currie 2012).
Thus, a distinct feature of modern-day Havana is coexistence of buildings constructed using different architectural styles, typical of different eras and trends. The
intertwining elements of baroque, classicist, neoclassicist, colonial and art déco architecture gained recognition of the UNESCO Committee, which already included
Havana's old town on the UNESCO World Heritage List in 1982 (Scarpaci 2000).

3.1.4 *"Special Period" (from 1990 until the present day)*

Already in 1990, as the Eastern bloc was falling, Fidel Castro proclaimed *El período
especial en tiempos de paz* (literally, the Special Period in times of peace). As part
of it, the Cuban government introduced a number of economic and institutional reforms, which were meant to enable continued functioning of a Communist country

amongst changed geopolitical conditions, where Cuba was cut off from assistance of the Soviet Union (dissolved in 1991). The new strategy focused mainly on development and promotion of the tourist sector, which had previously been considered unproductive by the revolutionary authorities. From that point onwards, international tourism became a powerful force shaping the spatial development of the Cuban capital city (Colantonio & Potter 2006). According to M. Coyula (2002), in the last decade of the 20th century, up to 80% of building investments were associated with the tourism sector. Therefore, the city was developing along the coast, where vast hotel compounds were created or modernised. During the Special Period, the activities of the authorities led to serious neglect of Havana's housing sector. New housing investments were abandoned in favour of hotels and the accompanying infrastructure, while the existing but degraded housing estates did not undergo the required renovations. Moreover, the appearance of the aforementioned luxurious leisure compounds on the coast led to the creation of mono-functional areas, which were not integrated with the remaining parts of the city (Hernández 2011).

Due to Cuba's economic stagnation, continuing since the downfall of the USSR, and negligible impact of transformations introduced as part of the Special Period's reforms, Havana's contemporary spatial and functional structure has remained almost unaltered since mid-20th century. What is more, the Cuban capital city did not undergo many processes typical of free market economies that took place in other Latin American cities in the 20th century. In several respects, the city remains to this day an original, one-of-a-kind relic of the past, which has been experiencing developmental stagnation since the 1950s. Petrification of the historical urban landscape is a symptom of this stagnation. First, Havana is characterised by low-rise buildings and its space lacks vertical residential or office condominiums, which are typical of metropolises such as Panama, Bogotá or Lima (Czerny 2014). Second, it does not have a clearly defined business centre outside the colonial old town. The registered offices of main banks and financial institutions are to this day located in the city's oldest district – *Habana Vieja*. On top of that, unlike other Latin American cities, contemporary Havana is not experiencing the urban sprawl phenomenon. Apart from a lack of investments in infrastructure, this is also a result of the embargo imposed on Cuba by the administration of the President of the United States John F. Kennedy (Koont 2007). It led to serious restrictions in access to fuel, which, in turn, limited the role of individual car transport in favour of public transport (Edge et al. 2006; Koont 2007). Thus, Havana did not experience intensive ribbon development along the city's main thoroughfares, which probably would have happened after the revolution in capitalist conditions.

The contemporary spatial and functional structure of the capital city of Cuba was described by G. Ponce Herrero (2007a), who also developed its structural model. First, the author divided the city into three main zones with different land use intensity. The first one – central zone (Sp. *zona central*) – overlaps with the spatially dense colonial centre, where institutional, commercial and business activities are concentrated (Ponce Herrero 2007a). The second one is the intermediate zone (Sp. *zona intermedia*), characterised by varied forms of land use. It combines the dominant residential function with commercial and service functions. There are also small craftsman workshops within its borders. The intermediate zone is characterised by high population density and degraded buildings. The third zone is a peripheral area (Sp. *periferia*), characterised by the lowest land use intensity. The structural model also includes four directions of development, referred to as corridors (Sp. *corredoros*): *Corredor Este* (east), *Corredor Oeste* (west), *Corredor Boyeros* and *Corredor Cotorro*.

With the structural model he developed, G. Ponce Herrero (2007a) referenced the sectoral model proposed by H. Hoyt in 1939 (Hoyt 1939). Apart from two concentric zones – the colonial centre (*zona central*) and the zone developed during the republican phase (*zona intermedia*), a number of other sectors with wedge layout can be determined in Havana. There are two sectors of light industry and workers' apartments, spreading towards the south-east and south-west, as well as a sector located between them, occupied exclusively by working class apartments, two sectors inhabited by the upper class, stretching out along the coast, and a sector inhabited by the middle class, spreading towards the south-west. Between these sectors, there are areas which are currently either not used or being successively developed. Although the model proposed by G. Ponce Herrero (2007a) was developed with the year 1997 in mind, it is identical to the situation from the mid-20th century presented by the same author (2007b).

The characteristics of contemporary spatial and functional structure of Havana reflect various periods of its development – from colonial through republican, pseudo-republican, post-revolutionary and the latest period initiated with the fall of the Eastern bloc. Each of those periods introduced more or less significant changes in the way the society managed its space. Taking into account the fact that urban agriculture was developing in the Cuban capital city when the urban tissue was already fully developed and established, the specificity of the spatial and functional structure of Havana is crucial to the distribution of urban gardens.

3.2 History of urban agriculture in Cuba and its contemporary institutional and legal framework[6]

Urban agriculture in Havana dates back to the 1990s. The fall of the Eastern bloc and dismantlement of the USSR stopped the financial and logistical support that not only the Cuban agricultural sector, but, in fact, the entire national economy strictly depended on (Chaplowe 1998; Altieri et al. 1999; Premat 2003; Buchmann 2009; Premat 2012). Moreover, socialist countries had been responsible for 85% of Cuba's trade (Altieri et al. 1999). As a result of outside assistance being cut off and imports being drastically reduced (by 75%), the country faced one of the most serious economic crises in its history (ibidem). Immediately after the fall of the USSR, the budget deficit reached 33%, and according to official data, in the years 1989–1993, GDP dropped by 35% (Rosset & Benjamin 1994 as quoted in Altieri et al. 1999). Due to the severe shortages in food supply, caused by the considerable reduction in trade with the largest economic partner – the USSR and by the trade embargo imposed by the United States – the closest and the most economically powerful neighbour, Cubans, especially city residents, were forced to use a different strategy for obtaining food. The food system had to be transformed so that cities, which were the largest population hubs, could increase their self-sufficiency and independence from import from rural areas, which up until that point had been focused on sugar production (Premat 2003) and could not keep up with supplying the entire island due to underdevelopment. The Special Period announced in 1990 entailed, apart from numerous economic reforms, a restrictive, large-scale savings programme and strict rationalisation of resources (Altieri et al. 1999; Buchmann 2009; Nelson et al. 2009; Premat 2012). One of the priorities aimed at mitigating the effects of the economic crisis was the development of a sustainable agriculture model that would include the principles of ecological and organic agriculture (Nelson et al. 2009). However, it should be noted that the urban agriculture movement, which had been almost non-existent in Cuba prior to 1989, was initially a spontaneous, grassroots initiative led by self-organising groups of residents, who viewed agricultural use of undeveloped urban land as a chance for deterring the perspective of famine, which was becoming increasingly realistic with every year. What was initiated by the urban population, which lacked experience in terms of agricultural techniques and only strived to meet its basic nutritional needs, was only later quickly and ho-listically institutionalised (Altieri et al. 1999; Murphy 1999). In September 1993, the Cuban government, as part of Law 142, introduced a new agrarian reform by dividing the majority of large, state-owned farms into small Basic Units of Coop-erative Production (*Unidades Básicas de Producción Cooperativa* – UBPC), used by workers (Murphy 1999; Novo & Murphy 2000; Buchmann 2009). During this time, 60% of state-owned farms, including those located within administrative bor-ders of cities, were transformed into UBPCs (Murphy 1999). A critical moment in the history of Cuban urban agriculture was the establishment of the Department of Urban Agriculture in the Ministry of Agriculture (MINAGRI) in 1994 (Altieri et al. 1999; Novo & Murphy 2000). Its main objective was making unused land available to those who wished to farm it and produce food in the city. Residents who were planning to establish an urban garden could apply for the right to use land (Herrera Sorzano 2015). This decentralised system allowed for quick transfer of land while keeping bureaucracy to the minimum. Moreover, in view of urgent nutritional needs of the residents, even private land within city borders was allocated to those who wished to farm it if that land was not used (Altieri et al. 1999).

At the same time, as a result of the agricultural reform, 121 local marketplaces opened in Cuba, allowing producers to sell their products directly to consumers instead of through the national distribution chain as it had been the case before (ibidem). What is more, the prices of agricultural products at marketplaces were independent from the authorities' decisions. Thus, interventionism was limited and the free market law of supply and demand was allowed to take its course. However, this freedom was only apparent since it applied exclusively to distribution of pro-duction surpluses. First and foremost, farmers and gardeners were bound by state contracts to supply the national food system; only after having done that were they able to sell the surpluses (ibidem).

Urban agriculture in Cuba is currently organised on a national, provincial and municipal level as part of a number of cooperative, private or semi-private na-tional structures (Novo & Murphy 2000; Koont 2007; Herrera Sorzano 2009) (see Figure 3.1). The highest level is the National Group of Urban Agriculture

state level

the National Group of Urban Agriculture

Urban Agriculture Provincial Division

provincial level

Urban Agriculture State Provincial Council

- Provincial Delegation (Subdelegations and Directorates)

Urban Agriculture Business Provincial Council

- Metropolitan Horticultural Company
- Clearing and construction company
- Various Crops Company
- El Trigal Marketing Company
- Bacuranao Livestock Company
- Company of Agricultural Projects
- Insurance Company, Services and Urban Agriculture Delivery

the Municipal Division of Urban Agriculture

municipal level

Municipal State Council

Municipal Business Council

State Sector

- State Farms
- *Autoconsumos Estatales*

Private Sector

- UBPC (Basic Units of Cooperative Production)
- CCS (Credit and Service Cooperatives)
- CPA (Agricultural Production Cooperatives)
- groups of individual farmers

28 subprogrammes

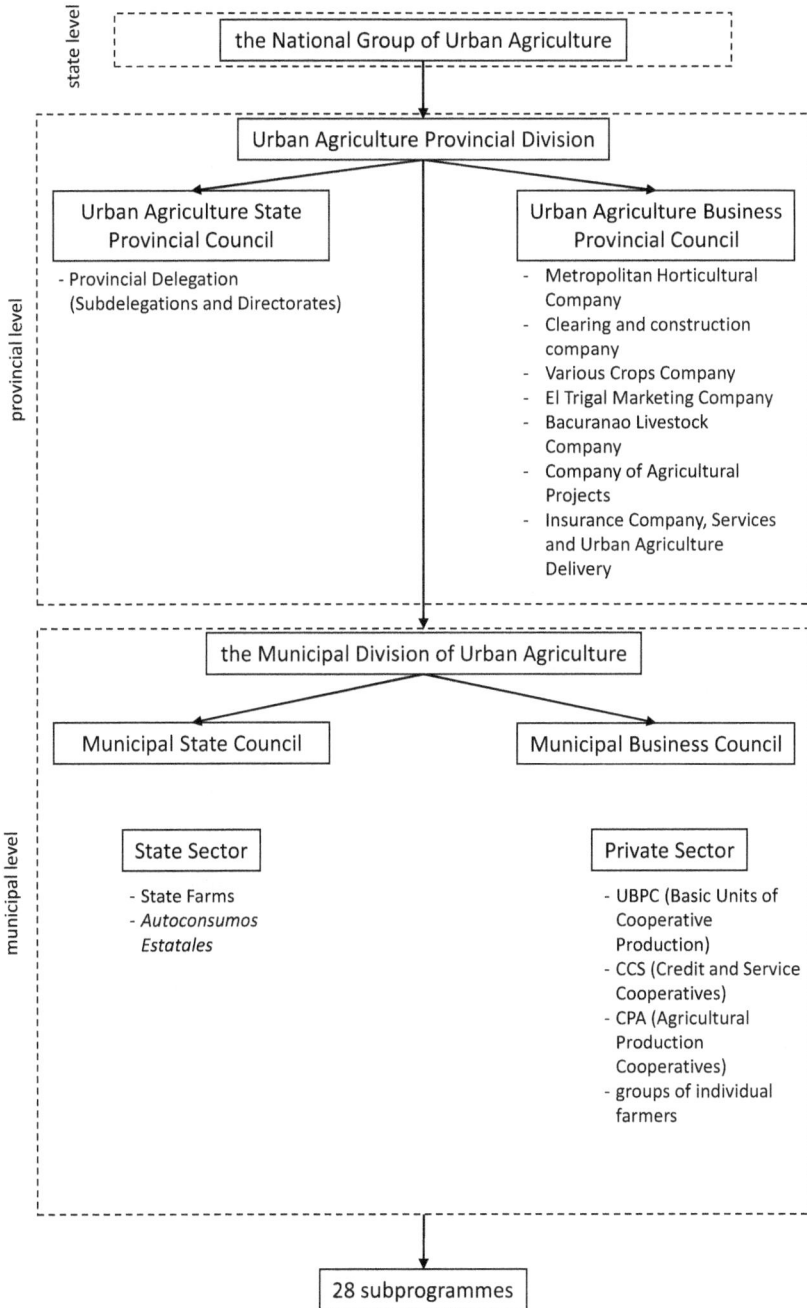

Figure 3.1 Organisational structures of urban agriculture in Cuba, divided into three institutional levels: national, provincial and municipal.

Source: Own study based on Herrera Sorzano 2009.

(Sp. *Grupo Nacional de Agricultura Urbana* – GNAU), which appoints a delegate at the provincial level. At the provincial level, in turn, there are two Provincial Councils for urban agriculture – the state council (Sp. *Consejo Estatal Provincial de la Agricultura Urbana*), divided into sub-delegations and directorates, and the business council (Sp. *Consejo Empresarial Provincial de la Agricultura Urbana*), which comprises numerous enterprises associated with the food production process, for example, the Metropolitan Horticultural Company (Sp. *Empresa Hortícola Metropolitana*), the *Bacuranao* Livestock Company (Sp. *Empresa Pecuaria Bacuranao*) or the Agricultural Insurance, Services and Delivery Company (Sp. *Empresa de Aseguramiento, Servicios y Suministros de la Agricultura Urbana*). Authorities at the municipal level include two Municipal Councils for urban agriculture – the state council (Sp. *Consejo Estatal Municipal de la Agricultura Urbana*) and the business council (Sp. *Consejo Empresarial Municipal de la Agricultura Urbana*), as well as the Municipal Division of Urban Agriculture, divided into the state sector, which brings together State Farms (Sp. *Fincas Estatales*) and *Autoconsmos Estatales*, as well as the private sector, which includes, among others, the previously discussed UBPC, Agricultural Production Cooperatives (Sp. *Cooperativas de Producción Agropecuarias* – CPA) or groups of individual farmers. Twenty-eight sub-programmes are implemented within the structures of the urban agriculture organisations described above. They are divided into three groups: crop sub-programmes (Sp. *subprogramas de cultivos*), including: vegetables and spices, medicinal plants and dried spices, decorative plants and flowers, tropical roots and tubers, cooking bananas, rice, maize and sorghum, beans, oilseeds, fruit, coffee as well as forestry and protected crops; livestock sub-programmes (Sp. *subprogramas pecuarios*), including cattle, poultry, sheep and goats, pigs, rabbits, aquaculture, apiculture; as well as supporting sub-programmes (Sp. *subprogramas de apoyo*), including land control, use and protection, organic matter, irrigation and drainage, seeds, animal feeds, small agroindustry, science, technology and training, the natural environment as well as marketing (Herrera Sorzano 2009). Structures of urban agriculture organisations are elaborate and it is not always possible to explicitly determine the hierarchical relations between individual authorities. Moreover, combined with the exceptionally complicated land ownership system in Cuba (Castañeda Abad et al. 2017), a detailed analysis of how those structures function has proved extremely difficult.

Urban agriculture in Cuban cities is not only highly institutionalised, but also strictly controlled in terms of distribution of the agricultural production resources and the use of organic agricultural practices (Nelson et al. 2009). Representatives of the Ministry of Agriculture – delegates at the provincial and municipal level – monitor urban gardens by checking whether their representatives apply appropriate production techniques (ibidem). Moreover, the state authorities also have a direct impact on the employment structure in the urban agriculture sector since they can directly delegate employees to work in individual gardens.

Government institutions also play an important role in the functioning of urban agriculture in Cuba because of the extensive gardener support system (Novo & Murphy 2000). In every city, a team of two to seven government experts in the field of agroecology offers assistance in terms of organic production techniques adapted to the needs of a given area (Altieri et al. 1999; Murphy 1999). What is more, gardeners can use specialised educational, technical and technological support within the scope of sustainable agriculture. This support is provided by non-governmental organisations, such as the Cuban Association of Organic Agriculture (Sp. *Asociación Cubana de Agricultura Orgánica* – ACAO), the National Association of Small Farmers (Sp. *Asociación Nacional de Agricultores Pequeños* – ANAP) or Seed Banks (Sp. *Casas de Semillas*), which also distribute the agricultural production resources (Altieri et al. 1999; Nelson et al. 2009; Castañeda Abad 2017). In Havana at the end of the 1990s, there were more than 400 horticultural clubs registered with the Ministry of Agriculture, where urban gardeners could share their agrotechnical knowledge and experience, as well as exchange agricultural production resources (Altieri et al. 1999; Murphy 1999).

The government's support and the introduced reforms have led to an increase in the number of urban gardens located on previously abandoned, unused and neglected land in Havana and other Cuban cities, while the strict control over agricultural practices ensured implementation of ecological production methods (Altieri et al. 1999; Nelson et al. 2009). However, this does not mean that gardeners always apply organic solutions in a fully informed manner. Based on research conducted in Havana, E. Nelson et al. (2009) demonstrate that a considerable portion of urban farmers choose organic techniques not voluntarily, but exclusively due to the institutional framework in place in Cuba, which prevents the use of conventional techniques. Although gardeners have an enthusiastic approach to ecological agriculture, the crop yield is the most important aspect to them and if they could access, for example, artificial fertilisers, they would opt to do so in order to increase their profits (ibidem).

3.3   Types of urban agriculture in Havana – a critical perspective

Urban agriculture in Havana is heterogeneous in terms of production volume and methods, species of plants cultivated and animals bred, as well as levels of management (Altieri et al. 1999). The diversity of urban gardens is mainly due to the aforementioned extensive institutional structures. Academic literature contains a number of types of Cuban urban gardens within a single typology that is reproduced by many authors, among others, M. Altieri et al. (1999), C. Murphy (1999), M. G. Novo and C. Murphy (2000), A. Herrera Sorzano (2009), S. Koont (2009), A. Premat (2005) and J. P. Díaz and P. Harris (2005). Although the following, commonly used division is meant to organise types of urban garden in terms of farming techniques and forms of organisation of the production process, it hardly provides a uniform and consistent typological procedure typical of geographical research. Nevertheless, discussing types proposed by such a big group of authors in renowned research on urban agriculture in Cuba will provide insight into the specificity of that country's urban gardens.

Due to their poor quality, anthropogenic soils, contaminated with heavy metals and construction materials, such as pieces of glass, concrete and plastic, are

Figure 3.2 Organopónico garden, where plants are grown on raised beds called *canteros*.
Source: Photograph by K. Górny, 2018.

typically unsuitable for farming. For that reason, an organic production method
called organoponics (Sp. *organopónico*) has been popularised in Cuban cities. It
entails growing plants on raised plant beds, that is, *canteros* (see Figure 3.2). They
are stabilised with borders made of various types of available materials, for example, roof tiles, concrete blocks or asbestos boards. *Canteros* are filled with a soil mix
with high compost content; the compost is produced on site or imported from rural
areas (Altieri et al. 1999; Murphy 1999). Non-composted organic waste available
on site is also frequently used (ibidem). Organoponics, which has been popularised
throughout the country since the 1990s, allows for agricultural use of undeveloped
plots of land, which otherwise could not be allocated for crop cultivation due to
the poor soil quality. Moreover, it allows for spaces covered with concrete to be
used as urban gardens (Murphy 1999). However, there are also examples of point-
less use of the organoponic method. There have been cases in Havana where *canteros* were built on fertile soil, where plants could be grown directly in the ground
(ibidem). The term *organopónico*, apart from referring to the described farming
technique, also provided the name for the first (and the most commonly described
in literature) type of urban gardens in Cuba that use the organoponic method. *Organopónico* gardens can be managed by national, cooperative as well as private
institutions and organisations (ibidem). Therefore, it is not a unit distinguished due
to the form of production organisation or the authority in charge for its operation.
Some authors also mention high-efficiency *organopónicos* (Sp. *organopónicos de
alto rendimiento*) (Díaz & Harris 2005; Premat 2005). This group is distinguished

through assessment of the production process efficiency. However, A. Viljoen and J. Howe (2005) emphasise that typically, such gardens are organised as part of the national structures. The term *organopónico* is frequently imprecisely equated with any urban gardens in Cuba due to the popularity and common use of organic agricultural practices in that country.

Another frequently differentiated type of urban agriculture in Cuba is the intensive garden (Sp. *huerto intensivo*). Similar to the *organopónico* type, it is a unit distinguished based on production techniques, regardless of its forms of organisation. The main difference between the two types is the fact that in an intensive garden, contrary to an *organopónico* garden, the plant beds do not have barriers (see Figure 3.3) and the soil is typically of sufficiently good quality for it to be mixed directly with additional organic material (Altieri et al. 1999; Koont 2009). As the name suggests, intensive gardens utilise intensive production methods. However, this is also the case for *organopónico* gardens. What is more, gardens of both types can be run by private, state or cooperative bodies (Murphy 1999).

The next type which appears in literature is called *autoconsumos estatales* or simply *autoconsumos* in Spanish. This is the first unit differentiated on the basis of the specifics of production organisation, not the methods applied. These gardens allocate the food they produce to supply canteens in various work establishments such as factories, military offices, schools, hospitals or care homes (Murphy 1999). The term *autoconsumos estatales* can therefore be translated as "self-supply state-owned gardens" or "intra-establishment gardens". Due to the purpose of the

Figure 3.3 A *huerto intensivo* garden type, where plants are grown directly in the ground.
Source: Photograph by K. Górny, 2018.

food produced in those gardens, they are also referred to as factory gardens or enterprise gardens (Altieri et al. 1999; Murphy 1999; Premat 2005). In *autoconsumos*, crops are first allocated to supplying the aforementioned work establishments
and then sold to workers at relatively low prices. Only after both distribution chan-
nels have been exhausted can the surpluses be sold directly to the local population (Novo & Murphy 2000). Thus, the degree to which *autoconsumos estatales*
are connected with the local food market varies depending on the level of commercial production, that is, production intended for sale. This group of gardens uses
both organoponics and crops grown directly in the soil, typical of intensive gardens.
A. Viljoen and J. Howe (2005) ascribe peripheral location to *autoconsumos estatales*. However, this attribute is not mentioned by other authors. Moreover, the
results of conducted field research also contradict this assumption since within the
limits of the designated area, characterised by dense architecture, there are also gardens which should be considered "self-supply state-owned gardens". *Autoconsumos
estatales* are also an institutional category at a municipal level since they are, apart
from State-Owned Agricultural Holdings, one of the units of the Municipal Division
of Urban Agriculture in the state sector (see Figure 3.1). The gardens visited during
the field research included those which, even though they are not unanimously officially referred to as *autoconsumo estatal*, have the features typical of this type of
gardens – they use the food they produce to supply work establishments.

Some authors also distinguish popular gardens (Sp. *huertos populares*), located on
private or state-owned land (Altieri et al. 1999; Murphy 1999; Díaz & Harris 2005).
They are typically abandoned or undeveloped parcels with an area ranging from
several square metres to several hectares, located in direct vicinity of the gardeners' households (see Figure 3.4). The right to use such land can be acquired for
free through the movement *Poder Popular* (literally People Power) – the smallest
organisational unit of the Cuban Government[7] (Altieri et al. 1999). The possibility of allocating undeveloped plots of land to crop cultivation was introduced in
1994, when the Department of Urban Agriculture in the Ministry of Agriculture
was established. Popular gardens are run privately by individual residents or local
organisations and horticultural cooperatives, with larger plots of land being typically divided into smaller parcels and divided among a larger group of residents
(Altieri et al. 1999; Murphy 1999; Díaz & Harris 2005). The food produced there is
typically intended to meet the needs of families of gardeners working on individual
plots of land. The remaining products are handed over to hospitals and care homes,
while surpluses are sold to the local population, as in the case of *autoconsumos
estatales* (Altieri et al. 1999). An important feature of popular gardens is the role
they play in supporting the population living in their immediate vicinity. Some of
the food produced there is supplied to secure the existence of those who need it the
most (senior citizens, the unemployed, etc.).

Another type of urban agriculture or a production unit in Cuba is referred to
as *parcelas y patios* by Spanish-speaking authors (e.g. Herrera Sorzano 2009) or
household gardens by English-speaking ones (e.g. Altieri et al. 1999). The former
name refers to the location of the gardens – on household plots or patios of buildings. The latter indicates the ownership of the gardens – private – and the allocation

Figure 3.4 A *huerto popular* garden located in a small gap between colonial buildings in the *Centro Habana* district.

Source: Photograph by K. Górny, 2018.

of the food produced – to meet the needs of individual households. Nevertheless, household gardens in Havana account for a relatively small portion of the overall area occupied by urban agriculture, even though they make up the most numerous group of gardens (Herrera Sorzano 2009).

The foregoing classification, which covers the most frequently encountered types of urban garden in Cuba, is not exhaustive and can be considered unclear and/ or incomplete. The presented garden types frequently overlap and do not always complement each other. Certain terms refer to methods and techniques of production, while others – to the organisation of the production process or exclusively to the method of product distribution. For instance, *organopónico* is a technique of production rather than an organisational category (Murphy 1999). The criterion for distinguishing this type of garden is exclusively the use of raised plant beds surrounded by borders. When plants are grown directly in the ground in a given garden and the plant beds, even raised ones, are without borders, we are dealing not with an *organopónico*, but with an intensive garden (Sp. *huerto intensivo*). What is more, both *organopónicos* and intensive gardens use intensive as well as organic methods of production, hence the distinction between these two types can be confusing. Nevertheless, both types of gardens are distinguished based on the production methods applied, which differentiates them from the remaining types, distinguished according to another criterion – organisation of the production process. These include *autoconsumos estatales*, popular gardens (Sp. *huertos populares*) or household gardens (Sp. *parcelas y patios*). All three groups can use either the organoponic method or the method of growing plants directly in the ground, as is the case in intensive gardens. The Cuban geographer A. Herrera Sorzano (2009) does not refer to the aforementioned groups of gardens as types, as is the case in English-language literature, but as production structures, which seems to be the most reasonable approach from the geographic point of view. *Organopónicos, huertos intensivos, autoconsumos estatales, huertos populares* and *parcelas y patios*, as production structures, may be organised as part of institutional structures – UBPC, CPA, CCS (Koont 2007; Herrera Sorzano 2009).

In this chapter, the terms *organopónico* and intensive garden will be used to indicate the production methods and techniques used, while *autoconsumos estatales*, popular gardens and household gardens will be treated as organisational categories, operating within the specified institutional structures (see Figure 3.1). Instead of classifying gardens as the individual types appearing in literature, individual gardens will be assigned attributes referring to the methods of production or organisation of the production process, including the method of product distribution. On the other hand, the broad term "urban gardens" or simply "gardens" will be used to refer to all urban agriculture sites, regardless of the applied production method and organisational form, even though its most precise use would be in relation to intensive gardens and popular gardens. Although this is a simplifying term, it is, at the same time, used most frequently in literature on the subject to indicate various forms of urban agriculture or horticulture in Cuba.

3.4   Distribution of agriculture within the urban space

Urban agriculture in Havana can be divided into two major groups, differentiated based on the location of the gardens within the urban space. The first group includes gardens located in densely developed areas. They should be classified as examples of intra-urban farming. It is those gardens that are analysed in detail in this chapter. The second group comprises farms located in suburban areas (which belong to the Havana province in administrative terms). Due to their links to the food system of the Cuban capital (they supply Havana residents with food and agricultural production resources), they are considered examples of peri-urban farming. Since they share more characteristics with agriculture typical of rural areas and their specific nature contrasts with the comprehensively analysed urban gardens located in densely developed areas, they are not included in this analysis for several reasons. First and foremost, farms located in the peripheries occupy a much larger space and are typically adjacent to one another, forming extensive areas used for agricultural purposes. Contrary to urban gardens, where intensive production methods are used and the crop structure is highly diversified, the agriculture in the city's outskirts is much more extensive, with frequent use of monocultures. Moreover, the origin of intra-urban farming in Havana differs from that of peri-urban farming. In the former case, this origin is secondary. Urban gardens in central districts of the Cuban capital city (e.g. *Plaza de la Revolución, Centro Habana, La Habana Vieja, Cerro* or *Diez de Octubre*) appeared in the aftermath of the economic crisis in the early 1990s.

They were founded on undeveloped plots of land, supplementing the established urban tissue. On the other hand, farms on the outskirts of Havana became incorporated into the city's borders in the process of its slow spatial expansion. Therefore, their origin is primary.

The following are the results of a detailed analysis of urban gardens in Havana, based on the field research carried out in 2018, preceded by an analysis of satellite and aerial images.

The detailed analysis covered 43 urban gardens located in the following districts of Havana: *Playa* (13 gardens), *Cerro* (9), *Plaza de la Revolución* (8), *Diez de Octubre* (5), *Centro Habana* (4), *San Miguel del Padrón* (2), *Habana Vieja* (1) and *Marianao* (1) (see Figure 3.5). Four abandoned plots of land, previously occupied by urban agriculture, were identified in the following districts: *Playa* (2), *Diez de Octubre* (1) and *Plaza de la Revolución* (1). The gardens are not distributed evenly within the limits of the city. One can point out two areas in which they are clearly concentrated. The first one is the vicinity of the Revolution Square (Sp. *Plaza de la Revolución*), where 13 gardens were identified within a radius of 1.5 km. The second area is located around the military airport – *Aerodromo Ciudad Libertad* (seven gardens within a radius of 1.5 km). Moreover, the gardens were also located on the borders of the research area, in direct vicinity of urban greenery areas in the south and south-east parts of the city (among others, the park on the *La Loma del Burro* hill or, for example, *Parque del pescado*) and along the coast in the vicinity of the street *Avenida 5ta* in the *Miramar* district, which is characterised by a lot of greenery and relatively large building quarters (Coyula Hamberg 2003; Colantonio & Potter 2006).

In areas with the highest building density, such as *El Vedado, Santos Suárez* and *Buenavista* districts, as well as the northern part of *Centro Habana*, no urban gardens were identified during the analysis of satellite images or during the field-work. This observation is in line with the results of research carried out by A. Premat (2009), who concluded that only small-scale crops, which include household gardens located, among others, on building patios or private balconies, are present in the strict centre of Havana. The small number of medium- and large-area gardens in densely developed districts results from shortage of space. According to A. Viljoen and J. Howe (2005), as population density and building density increases in Havana, the share of intra-urban farming decreases. On the other hand, the areas of concentration of urban gardens (in particular the areas near the Revolution Square and the *Aerodromo Ciudad Libertad* airport) are characterised by much lower density and less compact building placement. Such spatial distribution of urban agriculture results from the nature of Havana's urban layout. Thus, on the one hand, the city has a concise historical core, typical of Spanish hubs planned during the colonial period, which was densely developed for military and economic reasons. On the other hand, the outskirts of Havana, which developed along the main roads leading to the city's peripheries, are more dispersed and less densely developed (Viljoen & Howe 2005). In those areas, urban agriculture is typically concentrated near the main thoroughfares, such as *Avenida de la Independencia* (*Avenida de Rancho Boyeros*), *Avenida 5ta, Avenida 41* and *Calzada del Cerro* or in the vicinity

Figure 3.5 Distribution of urban agriculture in Havana, broken down by gardens using the organoponic method and intensive gardens. The map shows two main areas in which gardens are concentrated – the vicinity of the Revolution Square and the areas near the military airport, Aerodromo Ciudad Libertad.

Source: Górna and Górny 2020.

of urban greenery, industrial zones and new residential districts, where space is more readily available. A. Premat (2005) concludes that apart from the outskirts of Havana, *organopónico* gardens were concentrated in those parts of the city where building density was low. This is also confirmed by the results of research carried out for the purposes of this book.

A. Viljoen and J. Howe (2005) demonstrated that in the city outskirts, which can be considered the peripheral zone identified by G. Ponce Herrero (2007a), where buildings are dispersed, the area of urban gardens increases. However, the authors, like W. Castañeda Abad et al. (2017) in other studies analysing the entire La Habana province, were referring to both intra-urban and peri-urban farming. Taking into account only the gardens belonging to the former group, the authors' conclusions do not apply in this case. Among intra-urban gardens, those with the largest area are located in the city centre, in the vicinity of the Revolution Square. It is in the *Plaza de la Revolución* district, near the most prominent city square in Cuba, that the most important public buildings and government facilities are located, and the square itself has been a site of celebration of key national occasions since it was created. Despite its representative role, a considerable portion of space in the *Plaza de la Revolución* district is occupied by crops (see Figure 3.6). This situation could be considered unusual, given the fact that land in city centres is typically so valuable that it is not allocated to activity which does not produce calculable economic gains. Due to its low profitability, urban agriculture usually loses the fight for space

Figure 3.6 High-rise (left) and low-rise (right) modernist buildings in the vicinity of the Revolution Square (so-called Second Republican Centre) with vast swathes of greenery between buildings. Two urban gardens are visible in the photograph.

Source: Photograph by K. Górny, 2018.

with other, more lucrative forms of land use. High availability of spatial resources in the *Plaza de la Revolución* district is the direct result of modernist urban planning, commonly applied both at the end of the pseudo-republican period and after the Cuban revolution. According to Le Corbusier's vision contained in the Athens Charter (1933), vast swathes of greenery were intentionally left between buildings so that the apartment buildings appeared to be "growing" in a forest. Entire housing estates of large panel apartment buildings were erected in Havana in accordance with this concept, which was extremely popular (among others) in socialist countries. Since the beginning of the 1990s, it was those areas that became successively occupied by urban agriculture. Securing the rights to use unused public areas which were not allocated for urban development became even easier after the establishment of the Department of Urban Agriculture in 1994 (Altieri et al. 1999).

The situation is similar for the eight gardens located in the *Playa* district along one of the city's main thoroughfares – *Avenida 5ta*. Although contrary to the vicinity of the Revolution Square, this area is characterised by densely located villas and the spaces between the buildings are small; in this area, the presence of urban agriculture is also determined by the supply of gaps in the urban tissue. Before the Cuban revolution, *Playa*, in particular its east part called *Miramar*, was inhabited by the local elites. After 1959, those elites left behind numerous luxurious residences. Currently, *Miramar*, in particular the vicinity of *Avenida 5ta*, is known as the "embassy district". Urban gardens located within it take up the small spaces between individual mansions. On the one hand, their considerable number is due to the availability of land resources, similar to *Plaza de la Revolución*. On the other hand, it is due to the resident's needs for food. Since *Miramar* was inhabited by Havana's elites, it was intentionally deprived of commercial facilities, including grocery stores (Coyula Hamberg 2003; Currie 2012). Urban gardens, in response to the nutritional needs of the residents, were established here using free spaces (much smaller than in the vicinity) and "squeezed" in between the buildings.

Different historical models of spatial planning, which were applied in individual districts of Havana, affected the contemporary spatial layout of urban agriculture in the Cuban capital city. Uneven distribution of places of food production and distribution has serious consequences for the food security of the local population. The number of urban gardens in areas with high building density is insufficient to satisfy the demand. For this reason, many residents of districts such as *Centro Habana* or *Habana Vieja* have to go to other districts in order to purchase basic food products (Murphy 1999).

Factors which affected the distribution of urban gardens in Havana also include the administrative decisions of the authorities. Due to the high institutionalisation of urban agriculture, any actions aimed at local food production taken by residents are subject to strict state control. Therefore, establishing a garden in a specific location requires a permit from the competent administrative authorities. Thanks to it, the authorities are able to control the directions of development of urban agriculture in the Cuban capital. However, their actions are to some degree limited due to the specificity of the spatial and functional structure of the urban organism as well as the spatial distribution of plots of land available for development. Prominent actors who have a direct impact on the distribution of urban gardens in Havana are the residents themselves. After all, the beginnings of urban farming were primarily the result of their grassroots actions caused by their nutritional needs.

When discussing the features of distribution of urban gardens in the Cuban capital city, one should consider how the number of gardens changed within the designated research area in the years 2000–2020 (see Figure 3.7). The dynamic analysis was carried out based on satellite and aerial images in Google Earth Pro. The orthophotomaps used came from 2000, 2001, 2003, 2004, 2006 and 2008–2020. The results were extrapolated for the years for which no image was available (2002, 2005 and 2007).

Apart from the 43 gardens visited during the field research (in Figure 3.7, they were marked with numbers from 18 to 60), the analysis also includes 17 additional locations (see Figure 3.7, numbers from 1 to 17). Twelve of them are sites which no longer existed in 2018 (the year of the field research). Two other gardens had a hiatus in their operation, while the remaining three were not visited in the field for reasons independent from the author (the gardens were inaccessible to outsiders). Only 26 of the 60 gardens operated throughout the entire period covered by the dynamic analysis (2000–2020). They were sites located in the districts of *Playa* (7), *Cerro* (7), *Plaza* (6), *Díez Octubre* (3), *San Miguel* (1) and *Marianao* (1). Four gardens which operated in early 21st century were closed down permanently and the plots of land where they were located were built over. However, similar cases are rare, mainly due to the slow pace at which building density progresses in Havana, where unoccupied parcels remain a permanent component of the landscape. Much more frequently, gardens were abandoned and the land was transformed into wastelands (eight such cases were recorded in all of Havana). Among the liquidated or abandoned gardens, as many as eight were located in the *Playa* district, which is one of the most prestigious districts in Havana, with embassies, luxurious hotels and suites. This explains the liquidation of four gardens located in that district, which were allocated for development over time. At one of such sites, a luxurious suite with a swimming pool was constructed in place of a former *organopónico*. However, the prestigious status of the district does not explain the large number of abandoned gardens but rather stands in contrast to this fact. Deserted, overgrown plots of land left after those gardens have a negative effect on the landscape qualities of the housing estate, contrasting with the well-maintained household gardens.

The analysed sample includes two gardens which suspended their activity for three years (see Figure 3.7, No 10) and one year (see Figure 3.7, No 9). In both cases, during the analysis of the orthophotomaps from different years, a hiatus in crop cultivation was observed, followed by its later resumption. It is difficult to determine the reason for suspension of activity of the *organopónico*; however, it may result from restriction of funds or worker shortage. Similar causes may be behind the liquidation of two gardens, one after nine years (see Figure 3.7, No 13) and the other one after six years (see Figure. 3.7, No 14) from their establishment.

The overall number of gardens in individual years ranged from 36 to 50. The majority of the 60 analysed gardens (38) operated in 2000. However, during the first two decades of the 21st century, urban agriculture started successively appearing in

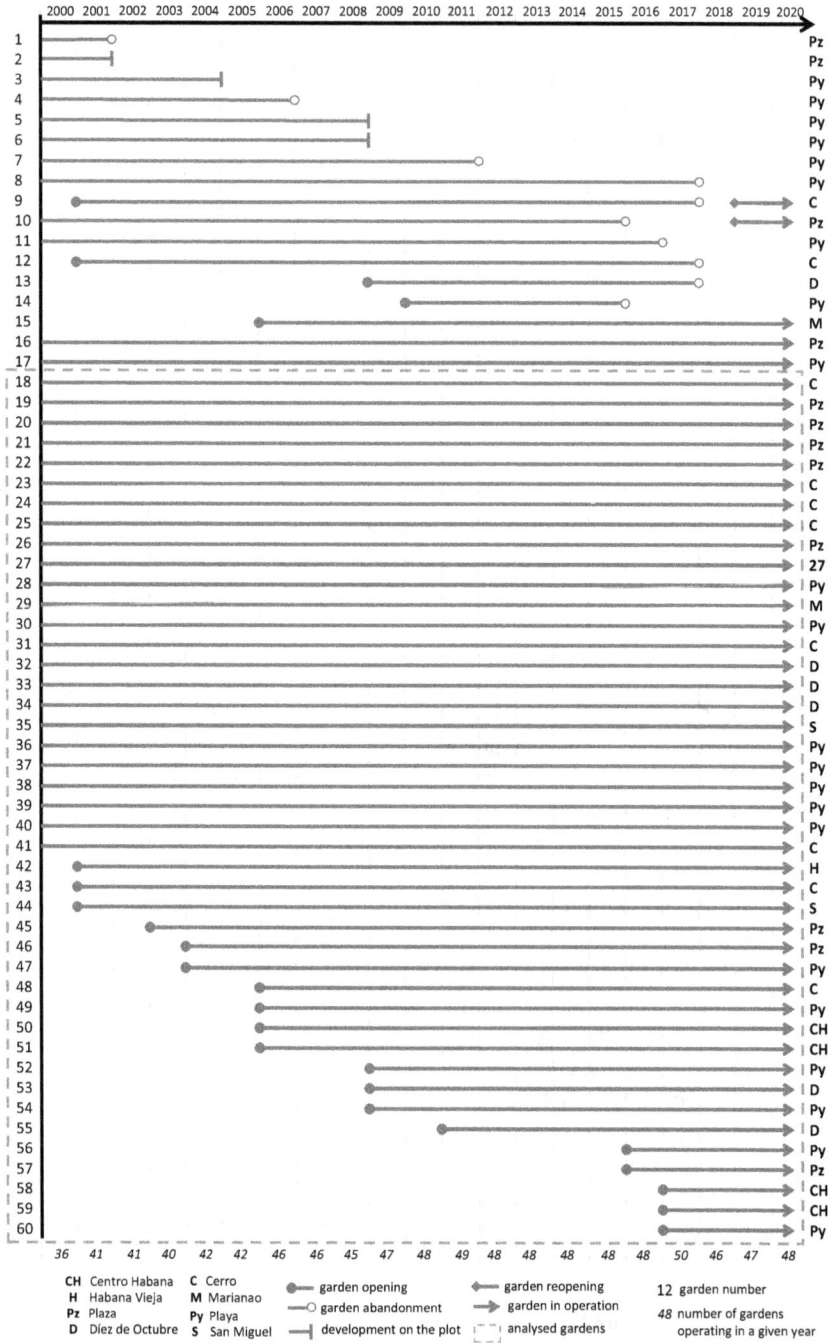

Figure 3.7 Urban gardens operating in Havana in 2000–2020.

Source: Own study.

new locations. From 2000 to 2010, 18 new gardens were established and six more of them appeared in the following decade, five of which date back to the last five years. While new gardens appeared in the research area, others were liquidated or abandoned. In the analysed sample of 60 gardens, only 12 were shut down permanently and only four plots of land occupied by them were allocated to development. The changes took place at the fastest pace in the *Playa* district, where seven new gardens were established and eight were shut down.

The foregoing analysis was updated with data for 2022, collected based on satellite and aerial images in Google Earth. Among the 60 gardens covered by the analysis, changes were only observed in two locations (see Figure 3.7, No 46 and No 10). In the case of garden No 46, no crops are visible in the image, which may indicate suspension of activity. Nevertheless, the options for determining this conclusively are limited due to the low image quality. In the case of garden No 10, a change in land use compared to 2020 is visible – an orchard is currently located in the place of ground cultivation.

The successive increase in the number of gardens in the research area in 2000–2020 (combined with a small decrease in 2017–2018) leads to a number of conclusions regarding the development of urban agriculture in Havana and development of the capital city itself. First, the fact that new gardens were established in the city space demonstrates sustained or increasing demand for locally produced food. The economic situation on the island has not improved considerably since the early 1990s. As a result, on the one hand, measures aimed at reducing the pressure on agriculture in rural areas are still needed; on the other hand, the residents themselves are experiencing issues associated with maintaining food security, so they opt for production to satisfy their own needs. Therefore, establishing new urban gardens and maintaining the operating ones is a response to Havana's food problems. Moreover, an important factor influencing the presence of urban agriculture in the capital city of Cuba was support for organic production in cities on the part of state authorities and legislation allowing for occupancy of wastelands to grow crops. The government's policy as part of the struggle against the aftermath of the economic crisis of the 1990s supported not only maintenance of existing agriculture within Havana's space, but also establishment of gardens in new locations.

Second, the fact that the space used for urban agriculture purposes had not been previously allocated to the other, more lucrative forms of land use (only four plots of land after liquidated gardens have been developed) demonstrates that no visible measures aimed at increasing building density are currently carried out in Havana. The city is not developing dynamically enough for it to need to allocate land occupied by urban gardens to construction of new buildings and housing estates. Moreover, agricultural activity is not being pushed out to areas more and more remote from the centre. Due to limited investments in buildings and the small number of extensive revitalisation projects, the city can be considered stagnant. This situation is certainly conducive to maintenance of urban agriculture within Havana's space, as otherwise it would be the first in line to be sacrificed in favour of building expansion.

3.5   Characteristics and functions of urban agriculture

3.5.1   Structural and production characteristics

Functions of Google Earth Pro enabled estimation of the area of each of the ana-lysed urban gardens. For the purposes of this analysis, both the total plot area and the crop area were taken into account. In the majority of cases, both values differ considerably. The total area of all visited gardens amounts to 323,035 m^2, whereas the total area of crops themselves within the limits of those gardens is smaller by half and amounts to 152,090 m^2. This is due to the fact that urban agriculture in Havana, apart from food production, also fulfils other functions – social or recrea-tional. The land around the crops is typically occupied by accompanying facilities, such as seed nurseries, sales points, markets, tool sheds and other technical build-ings, compost bins, water tanks and cafes. The analysed urban gardens in Havana also vary in terms of size. Their areas range from 290 m^2 to 95,425 m^2, with an av-erage area of 7,512 m^2. On the other hand, areas used directly for agricultural pro-duction purposes take up merely from 96.8 m^2 to 73,448 m^2, with an average area of 3,621 m^2. State farms are considerably larger than small gardens managed by horticultural cooperatives or groups of residents. It should also be emphasised that an important factor affecting the area of the gardens is access to land – in districts with higher building density, the area of parcels allocated to urban agriculture is much smaller than in the more sparsely developed districts. Similar conclusions on the basis of research carried out throughout the La Habana province were reached by W. Castañeda Abad et al. (2017).

Based on the calculated total area of gardens in Havana and the average effi-ciency of agricultural production in Cuba described in the literature on the subject, it is possible to estimate the probable production yield and thus the efficiency of urban agriculture in the Cuban capital city. M. A. Altieri et al. (1999) point out that according to official data, in gardens that use the organoponic method, the annual production potential in 1996 amounted to 16 kg/m^2, while in the so-called intensive gardens, it stood at 12 kg/m^2. However, authors indicate that in many provinces of the country, production efficiency in the same year exceeded 20 kg/m^2. They also provide the average value for both groups of gardens, amounting to 18 kg/m^2 per year. Other authors – A. Viljoen and J. Howe (2005) – attribute different production efficiency to individual types of gardens. Thus, in 2000, it amounted to 20 kg/m^2 per year for *organopónico*, 25 kg/m^2 per year for *organopónicos de alto rendimiento* and from 8 to 12 kg/m^2 per year for intensive gardens. They also provide values for popular gardens (from 8 to 12 kg/m^2) and *autoconsumos estatales* (0.6 kg/m^2). Low production efficiency of the last group of gardens is due to the fact that the authors, contrary to the assumptions adopted in this book, assign exclusively peripheral lo-cation to those gardens and thus a more extensive nature, where high crop yield is achieved through increasing the crop area, and not, as in the case of gardens located in the central parts of the city, intensive production methods. According to the data provided by A. Viljoen and J. Howe (2005), production efficiency in all groups identified by them clearly increased between 1996[8] and 2000. It grew by 17 kg/m^2

for *organopónico*, 13 kg/m^2 for *organopónico de alto rendimiento*, 6–11 kg/m^2 for popular and intensive gardens and 0.26 kg/m^2 for *autoconsumos estatales*. This shows that urban agriculture in Cuba experienced a significant improvement in productivity within merely four years. S. Rodríguez Castellón (2003) reached similar conclusions, although on the basis of different data. According to him, the production yield in Cuban gardens using the organoponic method increased from 1.5 kg/m^2 in 1994 to 25.8 kg/m^2 in 2001. This spectacular success is the effect of the authorities' coherent policy in terms of supporting local organic food production and decentralised supply system for production resources and trainings addressed to producers (Koont 2008). Nevertheless, S. Singh et al. (2008) also provide other data concerning the production yield in urban gardens in Cuba. The values they published vary from those presented above, and contrary to the former, they do not indicate increased productivity. According to S. Singh et al. (2008), in 2006, the average annual production yield for vegetables and herbs amounted to 18.44 kg/m^2 for organoponic gardens and 11.3 kg/m^2 for intensive gardens. Taking into account the lack of current data, the significant discrepancies in those published by the aforementioned authors and different understanding of types of urban agriculture in Cuba, precise calculation of the production yield in the gardens analysed in this chapter is considerably impeded. For illustrative purposes only, using the average value of 18 kg/m^2 stated by M. A. Altieri et al. (1999), the production yield in the 43 analysed gardens in Havana can be estimated at 2,488,172.4 kg per year, while the average annual yield of one garden – at 57,864.47 kg per year. Using the population size of Havana from 2018 (i.e. the year of field research), which amounted to 2,136,000, and based on the assumption that no significant changes have taken place in the crop yield in Cuban urban gardens (which should be considered improbable in spite of everything), the production yield per one resident supplied by urban gardens within the research area would amount to 1.16 kg/year. M. A. Altieri et al. (1999) estimated the total production yield supplied by urban agriculture (both intra-urban and peri-urban) to Havana's residents at 8,500,000 kg, which amounted to 3.94 kg per resident. Thus, using the estimated value of 1.16 kg/resident/year, one can assume that products from intra-urban gardens can account for up to one-third of this value.

The predominant types of products grown in Havana's gardens are fruits and vegetables. This is a universal characteristic of intra-urban farming worldwide. Since fruits and vegetables spoil easily, it is key to reduce the distance between the place of production and the place of distribution. In the case of Havana, the situation of the city's residents in the early 1990s is also significant. During the economic crisis, residents of big cities were the most severely affected by the reductions in fruit and vegetable supplies from abroad, which *summa summarum* led to the emergence of urban gardens in the capital. Currently, the plant grown in Havana the most frequently is lettuce, present in 24 gardens visited during the field research. It is relatively easy to grow since it does not have excessive requirements in terms of soil. According to the conducted interviews, this species is grown continuously (all year round), which ensures its constant supply. Other vegetables frequently grown in Havana include onion (in 15 gardens), spinach (ten gardens), tomatoes (nine gardens) and parsley (six gardens). The most frequently grown fruit species are bananas (six gardens), mangoes (three gardens) and papayas (two gardens). Fruit trees are frequently planted in the outer parts of the gardens as a hedge separating the fields from the streets and providing partial shade. In some locations, herbs such as basil, oregano or spearmint are also grown. The most frequently grown medicinal herb is aloe, present in ten gardens. In many instances, this species is also used as a decorative plant. The breakdown of plant production in individual gardens is varied. In 27 cases, more than three vegetable species were grown within a single plot of land. Moreover, different species, such as spinach and lettuce, were often present on a single plant bed (intercropping).

All analysed gardens are involved in plant production. Its domination is yet another characteristic feature of urban agriculture. However, one should also mention 12 gardens in which plant growing is accompanied by animal husbandry. The most predominant type of animals is poultry, bred in ten gardens. Hens or ducks typically move around freely within a fenced off plot or land or are kept in small, vertically arranged cages, which limits the space they occupy. Other animals bred in Havana are goats, rabbits and pigs. Breeding of the latter is not common in central districts of the city since Cuban legal regulations require pig houses to be located in the peripheries, at a safe distance from residential areas in order to reduce the risk of contaminating water resources (Novo & Murphy 2000). The structure of animal production is less varied than that of plant production. Only in four gardens was more than one animal species observed. However, this is due to the aforementioned fact that animal breeding is only an ancillary activity in Havana.

3.5.2   *Organisational and technical characteristics*

The most common production method identified during the fieldwork was organoponics, which is in line with the results of research carried out previously by other authors (Altieri et al. 1999; Díaz & Harris 2005). It is applied in 34 urban gardens, which makes up 79.1% of all visited sites. According to research by W. Castañeda Abad et al. (2017), there were 96 organoponic gardens in all of La Habana province.[9] The use of organoponics results from low soil quality in the city, as well as the tradition of organic production, which developed over the last three decades. Raised plant beds are typically arranged in distinct rows, which allow for *organopónico* gardens to be identified in satellite and aerial images. *Canteros*, which are filled with soil, are constructed using various materials available in a given moment. These include medium-sized rocks (see Figure 3.8), roof tiles or concrete slabs (see Figure 3.9).

However, the dominant material used to construct the majority of *canteros* is asbestos. Its omnipresence is concerning, taking into account the serious health implications for the consumers and producers. Contact with this material, in particular in crumbled form, when it releases toxic substances, can lead to diseases such as asbestosis, mesothelioma and a number of other cancers, including lung cancer, laryngeal cancer, ovarian cancer or cancer of the digestive tract (Epelman 1993; Tweedale & McCulloch 2004). Due to the material's resistance to high

Figure 3.8 Canteros made of rock fragments.
Source: Photograph by K. Górny, 2018.

Figure 3.9 Canteros made of rock concrete slabs.
Source: Photograph by K. Górny, 2018.

temperature, humidity and chemicals, it was commonly used in residential construction after World War II. When it turned out that it is extremely dangerous to human health and life, its production was banned or considerably limited, especially in European and North American countries. As a result, asbestos, which was still being produced, started being exported to other regions, such as Latin America (Epelman 1993). Since the end of the Cold War, the USSR was one of the largest manufacturers of this material; Cuba became the main recipient of its supplies, which served as a form of support as part of cooperation between countries of the Eastern bloc (ibidem). Moreover, a company called *la Fábrica de Asbesto-Cemento "Armando Mestre"*,[10] which produces asbestos boards, operates to this day in Cuban cities and has its branches, among others, in Havana and Santiago de Cuba (Jiménez et al. 1995). During field research, asbestos use in constructing plant beds was observed in 26 of the visited gardens, which makes up 76.5% of all gardens using the organoponic method. *Canteros* were built using cut-up corrugated (see Figure 3.10) and flat asbestos boards (see Figure 3.11), as well as asbestos pipes (see Figure 3.12). Use of asbestos in food production is highly hazardous for both producers and consumers. The positive effects of using organoponics, the main role of which is reducing the negative effect of soil pollution on the produced food, are therefore squandered. This shows the significant obstacles in making full use of the advantages of organic production methods in situations when the resources are limited. Poverty, and therefore also lack of sufficient means, makes it difficult for residents to purchase new materials (e.g. wooden boxes, boards), forcing them

Figure 3.10 Canteros made of corrugated asbestos boards.
Source: Photograph by K. Górny, 2018.

Figure 3.11 Canteros made of flat asbestos boards.

Source: Photograph by K. Górny, 2018.

Figure 3.12 Canteros made of asbestos pipes. The photograph also shows semi-permeable
sheets that offer rain and sunlight protection.

Source: Photograph by K. Górny, 2018.

to use those already in their possession. The situation is additionally aggravated by the approach to food production in cities promoted by the Communist authorities, which involves minimising resource wastage. In spite of the government's support aimed at Cuban gardens, intended to improve the efficiency of crops and ensure optimal conditions for the development of urban agriculture, since the 1990s, asbestos still has not been replaced by any safer material. Taking into account the fact that the majority of residents involved in food production are uneducated, and sometimes even lack experience in agriculture and derive knowledge on organic production methods from trainings organised by state administration units, it is the governmental experts, responsible for the quality of food produced in Cuban gardens, that should protect the residents from disastrous effects of asbestos use. This raises the question of whether the authorities are aware of its harmfulness and intentionally opt not to take adequate steps, ignoring the danger, or whether even national experts lack knowledge on the toxicity of this ubiquitous material.

While the issues of asbestos use in construction and the resulting health consequences for Cuban residents were already touched upon in academic literature a quarter of a century ago (Jiménez et al. 1995), its use in urban agriculture was omitted. Only C. Murphy (1999) lists asbestos as one of the materials used to build *canteros*. However, the author does not specify the scale of its use or the health consequences for its application in food production. The ubiquity of asbestos in urban gardens in Havana highlights the poor level of farming culture in Cuba and thus lack of awareness of its harmfulness on the part of residents and perhaps also the authorities. Therefore, conducting research aimed at assessing the impact of substances present in materials used to construct *canteros* on the quality of food grown in them as well as health of producers and consumers seems necessary and particularly urgent. It would allow for determining the severity of the issue and draw the decision-makers' attention to its effects.

In eight of the analysed gardens, plants were grown directly in the soil. Therefore, they can be considered intensive gardens (Sp. *huerto intensivo*) according to the types discussed above. Such a production method is used in Havana much less frequently due to the poor quality of the surface soil layer (Altieri 1999).

Four small orchards were also located in the analysed area. However, in all instances, they accompanied organoponic or intensive gardens, supplementing production of annual plants with tree-grown fruit. The largest one – a mango orchard with an area of 13,675 m² – was a part of a larger *Organopónico "Santovenia"* compound in the *Cerro* district.

One of the visited sites was occupied by a company involved in production and sale of seeds, *Empresa Productora y Comercializadora de Semillas*. Although farming was not its core activity, its plot of land featured a small crop field of 677 m².

A distinct feature of Cuban urban agriculture is use of intercropping (see Figure 3.13). Planting various species within a single field, and frequently even plant bed, was observed in all analysed gardens. This method entails a number of benefits. First, such a practice allows for regular supplies of many products, which is particularly desirable when it comes to providing food to cities. Intercropping

Figure 3.13 Intercropping of vegetables and herbs.
Source: Photograph by K. Górny, 2018.

also prevents soil exhaustion, which is a common issue for monocultures. Moreover, co-occurrence of many plant species is a natural method of pest eradication. An example of it is planting basil next to cucumbers or tomatoes. In the case of the former, the presence of basil reduces the occurrence of mildew,[11] and in the case of the latter, it deters mosquitoes and flies.

Intercropping in Havana is the main method of pest control. Instead of synthetic chemical pesticides, allelopathic plants are used. Their ethereal oils deter soil pests and reduce weed growth. In certain cases, allelopathic plants have an opposite effect. Such properties are exhibited by marigolds and nasturtiums, for example. Their scent attracts pests, especially nematodes or aphids, preventing them from infesting the surrounding crops. The marigold (*Tagetes L.*) was an allelopathic plant used in 15 gardens. It was typically planted on the edge of plant beds or in their immediate vicinity (see Figure 3.14).

In six gardens, semi-permeable protective sheets were also used to protect the crops from excessive sunlight or precipitation (see Figure 3.12). They help increase crop yield and allow for planting of sun-sensitive vegetables and herbs. In literature on the subject, this technique is called semi-protected cultivation (Koont 2009). It is typical of agriculture in the Tropics – for example, it is commonly used in Singapore.

Due to the fact that the use of artificial fertilisers is prohibited in Cuban cities (Altieri et al. 1999; Novo & Murphy 2000), to improve soil fertility, gardeners in Havana use animal and organic waste from households or neighbouring work

Figure 3.14 Marigolds used to fight off pests, planted on the edge of a cantero in Organopónico "Plaza".

Source: Photograph by K. Górny, 2018.

establishments. Compost was produced on site in 23 of the visited gardens. This
observation differs from the research results published by C. Murphy in 1999. According to the researcher, compost used in Havana's gardens was obtained exclusively from rural areas. This discrepancy is most likely the effect of changes which
have taken place over the last two decades. Gardeners, supported by a number of
state institutions, improved their competences and capabilities in terms of organic
farming, which allowed for local compost production. The dominant method of
composting is the so-called pile method, that is, collecting waste in a pile typically located next to a fence (see Figure 3.15). This method can also be described
as open-air composting. It is common throughout the island (Altieri et al. 1999).
A closed composting bin was used much less frequently. In two gardens, vermiculture was also used, that is, composting assisted by farming different species of
worms, typically redworms, also known as Californian red worms or composting
worms (*Eisenia foetida*) (ibidem).

The majority of analysed gardens are watered manually, with rainwater collected for this purpose in dedicated tanks. Simple drip irrigation systems were
only used in 11 instances (see Figure 3.16). This method is more beneficial due
to reduced usage of water, which evaporates quickly in the case of traditional irrigation. Moreover, it allows for regular watering of plants without the need for
manual work performed by gardeners. It allows them to save time and take up
other, non-agricultural activities without neglecting the garden. Since there are two

Figure 3.15 Open-air composting in Organopónico "La Sazon".
Source: Photograph by K. Górny, 2018.

Figure 3.16 Drip irrigation systems in Organopónico "Santovenia".
Source: Photograph by K. Górny, 2018.

seasons in Cuba (wet and dry), irrigation intensity in Havana varies depending on the time of year. During the rainy season, which lasts from the turn of May and June until November, when rainfall is abundant, artificial irrigation, used mainly during the dry season, is not required.

In spite of substantial support from national institutions in terms of supply of production resources (including Seed Banks), eight gardens had their own seed nurseries in the form of premises made of net stretched over wooden or metal frames (see Figure 3.17). The presence of such facilities allows for independent supply of seeds and seedlings, which reduces dependence on the prices on the local market for agricultural products and increases the garden's self-sufficiency. It should be emphasised that production was not mechanised in any of the visited locations. It relied exclusively on the work of human muscles, which is typical of agriculture in many cities in Latin America. The small area of the gardens and high share of concrete surfaces prevents introduction of large machinery. The tools used in Havana (garden spades, metal watering cans, etc.) were typically simple and easy to maintain. Use of local production resources (such as organic matter or seeds) and inexpensive equipment (frequently made of reused materials) leads to lower costs and reduced dependency of the gardeners on the variable economic conditions, which ultimately increases their self-sufficiency.

Active promotion of organic urban farming by the state authorities has led to the development of a specific culture of ecological production among the residents. During the interviews, respondents appeared extremely proud of the fact that their

Figure 3.17 Seed nursery in Organopónico INRE-I.

Source: Photograph by K. Górny, 2018.

farming practices were organic, while their products were fresh and free from con-tamination. The fact that they produce ecological food had a positive effect on their perception of their activity. Many respondents, after confirming that their gar-den uses the popular organoponic method, seemed to become more confident and started responding to questions outright enthusiastically. Nevertheless, according to E. Nelson et al. (2009), use of ecological agricultural practices does not always reflect the farmers' informed intentions; instead, it results from the rules imposed by the government. Although this information does not negate the positive effects of agroecological reforms introduced by the state, it proves that modern-day urban agriculture in Cuba was shaped by top-down decisions. The lack of awareness of the harmful effects of asbestos among the residents and the authorities' negligence in terms of measures aimed at its elimination from gardens in Havana demon-strate that there are material barriers to local and organic food production amongst the limited flow of information (including information on the latest achievements in agrotechnology), typical of countries which are economically and politically isolated.

3.5.3   *Product distribution and functions fulfilled*

Product distribution is an issue which is particularly important in the context of shaping the local food system. Designating places of distribution of crops and de-termining the path taken by food from the producer to the consumer contributes to better understanding of the role played by urban agriculture in the spatial and func-tional structure of every city, including Havana. Products from 19 of the visited gardens (44.1%) were intended for sale to nearby residents as well as restaurants and diplomatic missions in the vicinity. Therefore, in those instances, agriculture supplied the local food market. Products could be purchased at small sales points (Sp. *puntos de venta*), typically located within the boundaries of the gardens them-selves (see Figure 3.18) or at the markets (Sp. *agromercados*) in their immediate vicinity. Local sale of fruits and vegetables means the producers incur no transport costs. Moreover, it allows them to avoid the risk of unsold products going to waste since producers can adjust the volume of fruit and vegetable harvest to the daily demand (Murphy 1999). Moreover, thanks to the sales points being located on site, the producer can directly interact with the consumers, while the latter can see the method and place of production of the purchased goods. This is a privilege that the overwhelming majority of city residents all over the world do not have.

Certain sales points or markets offered local food, while others, apart from fruits and vegetables grown locally, also sold products from rural areas. Therefore, urban agriculture is incapable of satisfying the demand for agricultural products on its own, and although it is an important component of Havana's local food system, it needs to be supported by rural agriculture.

In 14 cases (32.6%), agricultural products were intended for supplying the fol-lowing institutions: the office of the Ministry of the Interior (Sp. *Ministerio del Interior*) (1), the Ministry of Agriculture (Sp. *Ministerio de la Agricultura*) (1), the *Villa Marista G-2* Department of National Security (Sp. *Departamento de*

Figure 3.18 A sales point next to a garden located in *Diez de Octubre* district.

Source: Photograph by K. Górny 2018.

Seguridad del Estado Villa Marista G-2) (1), the canteen of the Enrique José Varona University of Pedagogical Sciences (*Universidad de Ciencias Pedagógicas
Enrique José Varona*) (2), the "Santovenia" care house (Sp. *Asilo de ancianos Santovenia*) (1), schools (6), a childcare centre (1) and a home for disabled people
(Sp. *Hogar de Personas con Discapacidad*) (1). Food produced in gardens which
have the characteristics of *autoconsumo* gardens was therefore intended mainly
for internal use of the aforementioned institutions and was not available to a wider
group of consumers or was only available to a limited extent. In 1994, urban producers obtained the right to sell their products directly to consumers. However,
some of them were bound by national agreements, according to which they were
obliged, in the first place, to supply institutions such as schools, hospitals or nurseries and only then were they able to sell their surpluses. In one of the visited gardens, run privately by individual residents, the situation was reversed – products
were first intended for sale, while possible surpluses were supplied to a school
located in the immediate vicinity.

The ten remaining gardens (23.3%) are the gardens where interviews could not
be conducted. Therefore, no information was obtained on the product allocation.
Due to the fact that no sales points were observed within the limits of the aforementioned gardens, one should assume that the food they produce is intended to meet
the needs of people who work there and their families (thus the producer is also
the consumer) or handed over for sale at local markets. In both cases, the value and
supply chain is considerably shortened.

Apart from the link to the local food market, the analysed gardens are also linked to the local labour market since they offer employment opportunities to the city's residents. The number of employees in each garden ranged from one to 16 people, with an average number of four workers. They either took care of the crops or worked at the sales point. Due to the fact that information on employment was only collected in 19 facilities, the foregoing values can hardly be treated as representative for all analysed cases. An important finding of the conducted research is prevalence of men among garden employees. Only three out of the 21 respondents were women (14.3%). Similar conclusions regarding over-representation of men among people engaged in urban agriculture in Havana were reached by O. Plonska (2017). According to estimates of M. C. Cruz Hernández and R. S. Medina (2001), the share of women in food production in Havana may range from 10% to 15%. A. Premat (2005), on the other hand, emphasises that the role of women in urban agriculture in Cuba is underestimated and their activity is frequently not captured in official statistics. According to the researcher, women are particularly important actors in small-scale and self-supply gardens. Although their role is significant, it is much less public.

Urban agriculture in Havana should certainly be considered a multifunctional activity. This is what makes the selected case study so impactful and unique. The analysed gardens not only supply the local food market while simultaneously contributing to the residents' food security, but also offer employment in a country where unemployment in cities frequently entails a life below the poverty line. Moreover, urban agriculture in Havana also plays a spatial role, contributing to the development of continuous productive landscapes (Viljoen & Bohn 2014). The social function also appears to be particularly important. Due to the fact that urban agriculture in Havana began as a grassroots initiative of the residents, even nowadays, it plays an important role in strengthening bonds between people. Despite strong intervention and control of the state, it has retained its participatory nature. Two of the analysed gardens – *La Ceiba* and *UCPJEV* (analysed in detail in a later part of this chapter) – can be ascribed a prominent social function. In the former case, the plot of land features a cafe along with a dedicated recreational area, which serves as a meeting and integration place for the local population. In the latter case, the garden operated as part of a rehabilitation project. At other sites, the sales points themselves are frequently important places for residents to meet and talk. Another important feature of urban agriculture in Havana is also the involvement of the elderly, who are socially excluded in many cities of the world. Farms in Singapore are similar in this regard. Work and participation in daily life of the garden activates them and allows them to maintain permanent interpersonal connections. Apart from positive impact on the psychological and physical condition, work in the urban gardens gives senior citizens additional financial resources, which, in turn, protects them against economic exclusion. Moreover, according to S. Koont (2009), urban agriculture not only supports development of social bonds, but is also therapeutic for the people involved in the production process (Koont 2009). On the other hand, O. Plonska (2017, p. 70) takes it a step further and emphasises that for the residents, urban gardens in Havana are "their own small worlds [functioning] in a broader authoritarian, ideologised context". Although they are not separated from the top-down policy and control of the authorities, they do provide "moments of freedom". Moreover, the author emphasises that residents working in the gardens are strongly bonded with them – not only physically or economically, but also emotionally and spiritually due to being involved in their activity.

An important role of the visited gardens, resulting somewhat from the involvement of the authorities, is the informative/educational function. According to the objectives of the urban agriculture system created by national institutions, it should operate based on the resident's ecological knowledge. For this reason, delegates of the authorities train producers in organic production techniques. However, it is difficult to precisely determine the scope of implementation of informational activities addressed to residents, who are not so much producers as consumers. Information boards explaining the techniques of food production and healthy lifestyle, located within a garden, were only found in one instance. In other cases, the educational function of the gardens appears to be limited.

3.6   Selected examples of urban gardens

The following four case studies of urban gardens in Havana illustrate different approaches to food production in that city, both in terms of the methods applied and in terms of the work organisation. The presented gardens fulfil different functions and play different roles in the food system of the Cuban capital. The purpose of compiling four different examples is to demonstrate the heterogeneity of urban agriculture in Havana. The diagrams accompanying the descriptions show the spatial layout of all facilities located within individual gardens.

3.6.1   *La Ceiba Proyecto Comnit. Agroecológico*

La Ceiba Proyecto Comnit. Agroecológico is a garden located in the industrial district of *Habana Vieja* in *Calle Diaria*. The following facilities are situated in its vicinity: a school – *Escuela Emilio Nuñes Tallapiedra*; a baseball field, a transmission tower – *Termoeléctrica de Tallapiedra*; a harbour – *Muelle Osvaldo Sánchez*; a railway station – *Estación Cristina*; as well as a fort – *Castillo de Atarés*. The garden is surrounded by low-rise residential and industrial buildings. It is the only site of this type in *Habana Vieja*.

The interview was conducted with the manager of the garden, who is a physics teacher by profession and had decided to retrain and take up work in food production. This demonstrates another, very distinct characteristic of urban agriculture in Cuba, which is involvement of people who have neither agricultural education nor experience in this field. Moreover, working in the garden is very often an additional occupation for the gardeners, not their main source of income (Novo & Murphy 2000). The fact that the person responsible for managing the site has tertiary education leads to a twofold conclusion. First, it may indicate that the authorities have high requirements as regards to the gardener's competences, since the garden operates as part of the national project *La Ceiba Proyecto Commnit Agroecológico* financed

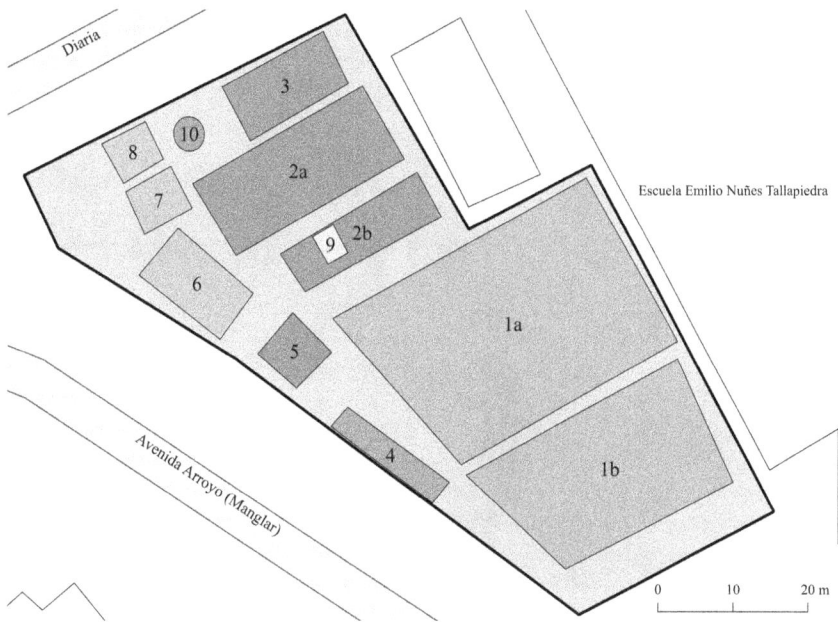

Figure 3.19 A diagram of the La Ceiba urban garden (1a and b – cultivation of vegetables directly in the soil; 2a and b – cultivation of decorative plants and herbs;
3 – semi-protected cultivation of decorative plants; 4 – a closed compost bin;
5 – cages with animals; 6 – farm buildings; 7 – cafe; 8 – sales point; 9 – seed
nursery; 10 – pond).

Source: Górna and Górny 2020.

with funds from the Ministry of Agriculture. Second, it can also mean that people
with high qualifications are unable to find a job that is in line with their profession,
so they become involved in urban agriculture.

According to the respondent, the garden was established in 2012. However, satellite and aerial images indicate that it had existed before, at least since 2001. The
date indicated by the respondent is probably the date of the official project launch.
Before the garden was established, the parcel, owned by the Cuban state treasury,
was abandoned and served as an illegal landfill. The area of the garden is 2,985 m²,
but the crops only take up 1,490 m², that is, approximately 50% of the area of the
entire parcel (see Figure 3.19).

La Ceiba can be considered an intensive garden (Sp. *huerto intensivo*), since the
plants are grown directly in the soil (see Figure 3.20). Importantly, as in the majority of the analysed cases, intercropping is used in this garden. It also combines plant
and animal production and is divided into the following sections: fruits and vegetables, decorative plants, aquaculture and animal husbandry. Individual areas of the
garden were presented in Figure 3.19. Within the plot of land, next to the cages with
hens and rabbits, there is a compost bin made of concrete slabs. Apart from plant
waste, animal waste is added to the compost. It is also one of two gardens that use

Figure 3.20 La Ceiba. Vegetables grown directly in the soil. Marigolds (*Tagetes* L.) have
been planted on the plant beds.

Source: Photograph by K. Górny 2018.

vermiculture. The compost bin is covered with egg packaging and a sheet protecting
it from excessive sunlight is suspended above it. The garden also features a small
reservoir in which ornamental fish are bred. Their faeces are used for compost production. Moreover, a small seed nursery ensures a continuous supply of seeds and
seedlings. Self-supply of production resources such as organic matter and seeds
reduces the need to purchase those products, which contributes to the garden's
self-sufficiency. The applied irrigation method is manual watering. No drip irrigation system has been installed on the field. What is more, similar to the majority of
gardens in Havana, no synthetic pesticides are used in *La Ceiba*. Instead, marigolds
(*Tagetes* L.) are planted on the edges of plant beds (see Figure 3.20).

   All of the food produced is intended for sale. The garden is adjacent to a school;
however, even production surpluses are not provided to that facility. The sales
point is located next to the street in the front part of the plot of land and is accessible to the public. According to the respondent, the prices are much lower than in
organopónicos in central parts of the city, which makes the products affordable for
the poorer population living nearby. Apart from the manager, the garden employs
six people. Two of them are responsible for growing fruits and vegetables, three are
in charge of decorative plants, while one deals with animal breeding. All employees live in the immediate vicinity.

   An important feature of the garden is the fact that apart from food production, it
fulfils a recreational function. First and foremost, there is a small cafe in the garden,

Figure 3.21 The sales point next to the entrance to the garden and a cafe for nearby residents.
Source: Photograph by K. Górny 2018.

which serves, among others, freshly pressed juices (see Figure 3.21). It is also a meeting place for the nearby residents, which strengthens their social bonds. This dedicated space, equipped with a television set, is used to organise various cyclical events, such as watching sports games or evening dances. The actors involved in the garden's operation – both employees and consumers – know each other and regularly spend time together, which is conducive to developing bonds within the local community.

3.6.2   *Organopónico "Santovenia"*

Organopónico "Santovenia" is located in *Calzada del Cerro*, the largest thoroughfare in the *Cerro* district. It is directly adjacent to a care home for the elderly, *Asilo de ancianos Santovenia*, run by a convent of *Hermanitas de los Ancianos Desamparados* (literally Sisters for Abandoned Elderly People). The garden is located in an area characterised by increased car and pedestrian traffic, surrounded by medium-rise industrial, residential and commercial and residential buildings. Near the parcel, there is a small petrol station, while across the street, there is another, much smaller care home and a rum factory – *Fábrica de Ron Legendario*. In the vicinity, there is also a hospital – *Salvador Allende (Covadonga)*, a company producing seeds – *Empresa Productora y Comercializadora de Semillas*, as well as a baseball field – *Estadio Latinoamericano*.

The interview in *Organopónico "Santovenia"* was conducted with one of the gardeners. He was accompanied by another employee, who was not the main interlocutor, although he did provide additional information. According to the respondent, the garden was established in 1998, which is in line with the analysis of

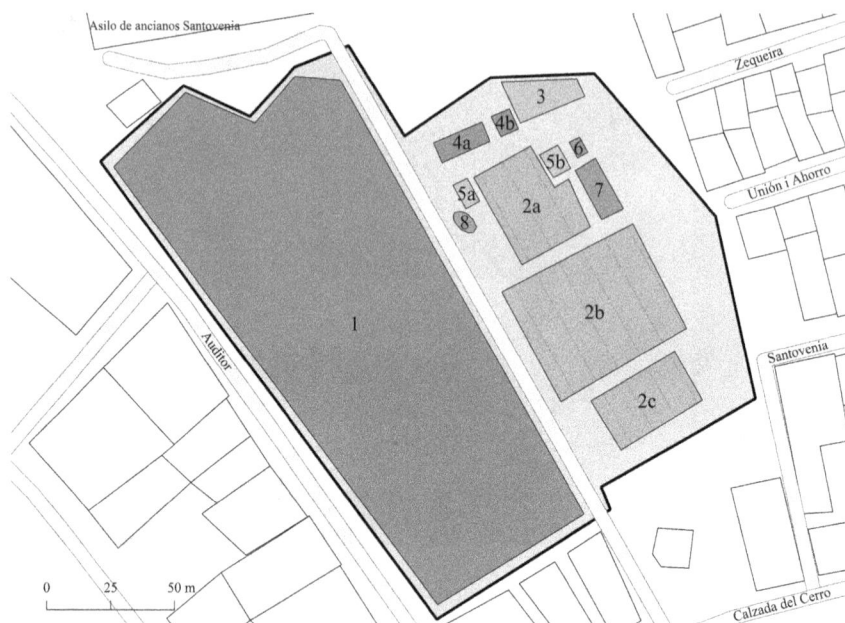

Figure 3.22 A diagram of *Organopónico "Santovenia"* (1 – a mango orchard; 2a and
c – organoponic cultivation of vegetables and herbs; 2b – semi-protected organoponic cultivation of vegetables and herbs; 3 – fenced off goat breeding; 4a
and b – cages with animals; 5a and b – farm buildings; 6 – abandoned greenhouse; 7 – aloe cultivation; 8 – compost bin).

Source: Górna and Górny 2020.

satellite and aerial images. The area is owned by the State Treasury and rented by
the convent. Similar to *La Ceiba*, the land had been unused before the founding of
the garden. The garden diagram is presented in Figure 3.22.

An internal road divides the area into two separate plots of land. One of
them is a mango orchard with an area of 13,675 m², while the other one, taking up 10,190 m², is a garden that combines several production systems (see Figure 3.22). *Organopónico "Santovenia"* differs from the previously discussed *La
Ceiba* primarily in terms of the applied production methods and allocation of the
products. As the name indicates, plants (vegetables, herbs and medicinal and decorative plants) are grown mainly in raised plant beds surrounded by borders (see
Figure 3.23). Organoponic crops take up a total of 3,767 m². *Canteros* are built
from fragments of corrugated asbestos boards and arranged into two groups of
parallel rows. Semi-permeable sheets, protecting them from sunlight, are stretched
out above them. The garden uses intercropping; however, the use of allelopathic
plants was not observed. Within the plot of land, there is also a small field, where
aloe is grown directly in the soil. Drip irrigation systems are installed next to the
organoponic crops (see Figures 3.16 and 3.23); however, in other parts of the garden, manual watering is used.

Figure 3.23 Partly shaded organoponic crops. The photograph shows a drip irrigation system and *canteros* made of asbestos.

Source: Photograph by K. Górny 2018.

According to the respondent, the parcel is used for breeding goats, hens and rabbits; however, only goats were observed during the fieldwork. Other animals were probably kept in the small building situated near the entrance. The garden produces compost, and similar to most visited locations, it uses an open-air compost bin for this purpose. Behind one of the farm buildings, in the most neglected part of the plot, there is a destroyed, unused greenhouse. During the fieldworks, the goats were seen grazing next to piled up waste, such as old beverage cans, bricks or plastic bags. The fact that the garden was neglected can be explained by an insufficient number of workers. In spite of the large area (it is one of the largest of the analysed gardens), only two people were employed there. By comparison, in *La Ceiba*, which is nearly eight times smaller than *Organopónico "Santovenia"*, the number of employees is more than three times higher.

The garden has features typical of an *autoconsumo*. The entire yield is used to supply the nearby *"Santovenia"* care home. According to the respondent, the mangoes from the orchard are also used to meet the needs of the convent's wards. *Organopónico "Santovenia"* is linked to the local food market to a limited extent since its products are not intended for sale. Instead, together with the *"Santovenia"* care home, it creates a small, closed-loop food system, where the place of production is directly connected with the place of consumption.

Organopónico "Santovenia" is also an interesting example of a garden which, according to the typology presented in this chapter, should be classified under two different categories. In terms of production methods, it should be considered an *organopónico* (as reflected by its name). On the other hand, taking into account the product distribution model, it shows characteristics of *autoconsumos*, although it is not officially described using this term. The case of *Organopónico "Santovenia"* demonstrates the aforementioned inconsistency of types adopted in literature on the subject.

3.6.3   *Organopónico INRE-I*

Organopónico INRE-I was established in 1991 and was the first *organopónico* founded in Havana. The garden is located in *Avenida 5ta* in the *Miramar* district, known as the "embassy district". In immediate vicinity of the garden, there are diplomatic missions of the People's Republic of China, Saudi Arabia, South Africa, Bolivia, Turkey, Mozambique, Italy and Ukraine. There are also two real estate agencies nearby: *Habana Palace* and *Edificio Montecarlo Palace*. The garden operates as part of the REDES project: *Proyecto Seguridad Alimentaria local en las provincias de la Habana, Artemisa y Mayabeque a través del fortalecimiento de la Agricultura Urbana I Suburbana y sus Redes de Servicios* (literally Local food security project in Havana, Artemisa and Mayabeque provinces through support of urban and peri-urban agriculture and their service networks). Its activity is also financed with European Union funds as part of the *Seguridad Alimentaria* (En. Food Security) project. *Organopónico INRE-I* occupies a state-owned plot of land and is under continuous supervision of the Ministry of Agriculture. The interview was conducted with a sales point employee. Additional information was also obtained from the manager of the garden.

   Organopónico INRE-I occupies an area of 4,730 m², of which 3,340 m², that is, almost 70%, is allocated to crop cultivation. This efficient use of space is a unique feature of the garden (see Figure 3.24). It results from its location in an area with high building density, which creates the need to use every square metre of space efficiently. The garden uses the organoponic method and the raised plant beds are arranged in two segments of parallel rows. It is one of few gardens in Havana where asbestos boards are not used to build *canteros* (see Figure 3.25). Instead, plant beds are sectioned off with concrete slabs. Production of *Organopónico INRE-I* is limited exclusively to vegetables, fruits and herbs. Therefore, there are no animals within its area. This can be due to the garden's location – in one of the most representative parts of the city. Breeding animals could result in unpleasant smells and noise, which is undesirable in a highly prestigious district.

   Some plant beds, where plants sensitive to excessive sunlight are grown, are covered with semi-permeable sheets. The composting method used here is open-air composting, just like in *Organopónico "Santovenia"*. However, it is supported by vermiculture, as in the case of *La Ceiba*. Marigolds (*Tagetes* L.) planted on the edges of *canteros* (see Figure 3.25) are used for pest protection. The garden also makes use of drip irrigation, which allows for reduction of water loss. The parcel features a seed nursery of 210 m² (see Figure 3.17). The scale of seed production is therefore larger than in the remaining visited gardens, where seed nurseries occupy at most 2 m².

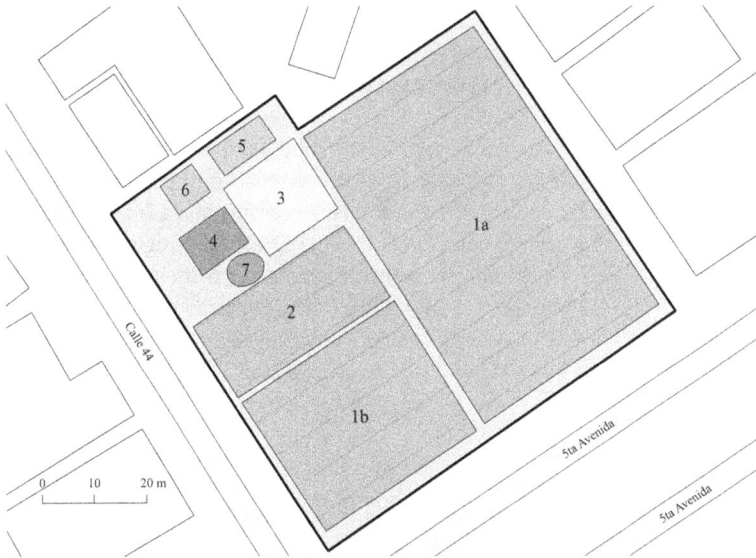

Figure 3.24 A diagram of the oldest organopónico in Havana – Organopónico INRE-I (1a and b – organoponic cultivation of vegetables and herbs; 2 – semi-protected organoponic cultivation; 3 – seed nursery; 4 – cultivation of decorative plants; 5 – farm buildings; 6 – sales point; 7 – compost bin).

Source: Own study on the basis of field research.

Figure 3.25 Organoponic cultivation of vegetables and herbs in Organopónico INRE-I. They are grown in *canteros* built from concrete slabs. Marigolds are planted on the edges of the plant beds.

Source: Photograph by K. Górny 2018.

Production in *Organopónico INRE-I* is fully commercialised. Vegetables, fruits and herbs are sold at the sales point located near the entrance. According to the respondent, although the store is open seven days a week, there is always a queue. Indeed, the interview was constantly interrupted by customers buying fruits and vegetables. *Organopónico INRE-I* is well known in the area and is frequently visited not only by permanent residents of the district, but also by employees of nearby embassies, consulates and restaurants. As emphasised by the respondent, non-Spanish-speaking employees of the Chinese Embassy regularly arrive with a shopping list and read it using a mobile translator on their smartphones. *Organopónico INRE-I* is an example of a garden firmly entrenched in the local food market. All of its yield is intended for sale, providing the people living and working nearby with a continuous supply of fresh food. The facility is popular in the vicinity due to its nearly 30-year-long uninterrupted production of food for the local market.

Organopónico INRE-I also fulfils and educational function. In front of the entrance to the sales point, there are information boards with content dedicated to food security and healthy nutrition. Some of the materials contained slogans such as "*¡Cuide su salud! ¡Consuma vegetales!*" ("Take care of your health! Eat vegetables!").

3.6.4   *UCPEJV Organopónico Dirección Producción VRES*

UCPEJV Organopónico Dirección Producción VRES is located within the compound of the "Enrique José Varona" University of Pedagogical Sciences (Sp. *Universidad de Ciencias Pedagógicas "Enrique José Varona"*) in the *Marianao* district, near the military airport *Aerodromo Ciudad Libertad*. A distinct feature of the garden is the fact that it operates as part of a youth rehabilitation project. The interview was conducted with a university employee (with a doctoral degree), who is also the manager of the aforementioned project.

The garden occupies 2,340 m², although only an area of 990 m² is allocated to crops. This is due to the fact that the production function is secondary in its case and high crop yield is not a priority. As the name suggests, the garden mainly makes use of the organoponic method, although certain plants, such as aloe, are grown directly in the soil. Rows of plant beds are arranged in two sectors (see Figure 3.26). The *canteros* are made of corrugated asbestos boards (see Figure 3.27) as well as corroded steel sheets. The crops are dominated by herbs and medicinal plants. However, lettuce is also grown on some plant beds. Edible plants are used to satisfy the needs of project participants, while their surpluses are handed over to the University canteen. Medicinal plants, on the other hand, are used in research on medicine production carried out at the University. According to the respondent, bananas (in particular *plátanos*, i.e. plantains) are also grown for this purpose.

The food produced in the garden is not intended for sale. Compost production uses organic waste from the nearby University canteen. During field observations, it was noted that the food leftovers included meat, which should not be used in the composting process. This demonstrates lack of adequate knowledge of gardening practices among people involved in the garden's operations.

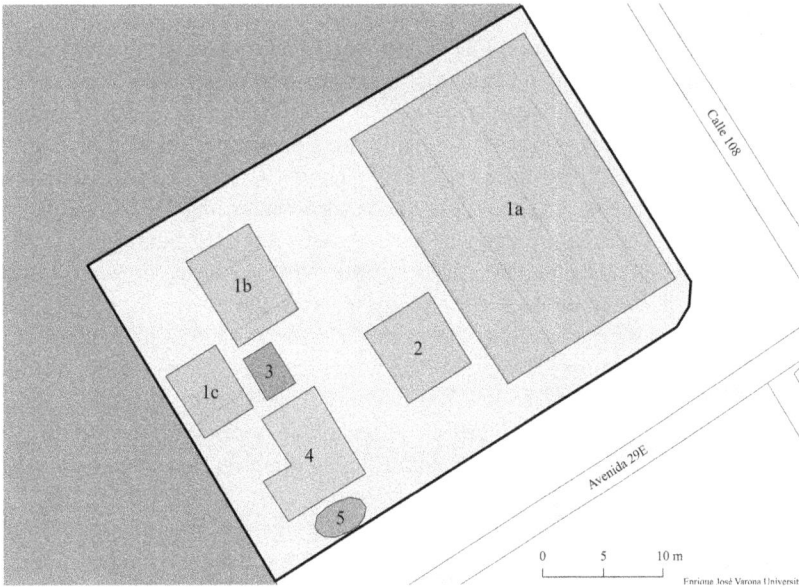

Figure 3.26 A diagram of UCPEJV Organopónico Dirección Producción VRES (1a, b and
c – organoponic cultivations of vegetables and herbs; 2 – cultivations of aloe;
3 – cultivation of plantains; 4 – farm building; 5 – compost bin).

Source: Own study on the basis of field research.

Figure 3.27 Rows of *canteros* made of corrugated asbestos boards. The photographs feature
marigolds planted within the plant beds as well as a drip irrigation system.

Source: Photograph by K. Górny 2018.

There is a drip irrigation system installed in *UCPEJV Organopónico Dirección Producción VRES* (see Figure 3.27). However, semi-protected cultivation is not used. It is not necessary due to the fact that the surrounding trees create a lot of shade on the parcel.

The garden operates as part of a rehabilitation project for the youth, described by the respondent in Spanish as *jovenes problemáticos* (literally troubled youth). Learning about organic production methods together, caring for the garden and being involved in mutual interactions are measures aimed at helping them adjust to life in the society. During the field observations, seven teenagers were working in the garden. The garden does not employ workers and all works are performed by the project participants under the manager's supervision. Although *UCPEJV Organopónico Dirección Producción VRES* does not supply the local food market, it fulfils other important functions. It is the only visited garden that contributes to scientific research while simultaneously conducting social activity. While many other gardens indirectly ensure community integration, in no other case was the social function given priority.

The presented case studies demonstrate the heterogeneity of urban agriculture in Havana. The visited urban gardens differ in terms of production methods, level of technological and technical development, organisation of the production process, spatial development as well as allocation of the food produced. Three of the discussed cases, which can be classified as examples of a single type – *organopónico: Organopónico INRE-I, Organopónico "Santovenia"* and *Organopónico VRES*, turn out to differ in terms of their links to Havana's food system and organisation of the production process. On the one hand, *Organopónico INRE-I* is an example of a garden with fully commercialised production. All fruits and vegetables are sold not only to the local population, but also nearby restaurants and embassies. On the other hand, *Organopónico "Santovenia"* is a garden whose products are intended exclusively for supplying the neighbouring convent and care home. *Organopónico VRES* is similar, but in its case, the products are intended to meet the needs of the project participants and the university, while the social function is dominant. *La Ceiba*, in turn, utilises production methods typical of an intensive garden, and similar to *Organopónico INRE-I*, it supplies the local food market. Although the garden also fulfils a social function, it can only be described as secondary. The presented diversity of gardens indicates the need to systematise types of urban agriculture. Production methods and forms of organisation should not be treated as separate and mutually exclusive types, but rather as characteristics which can be taken into account when developing a typology.

3.7   Urban agriculture in Havana – a summary

The conclusions drawn from the analysis performed are both cognitive and methodical. On the one hand, they concern the internal characteristics of urban gardens in Havana and their place in the city's spatial and functional structure; on the other hand, they provide practical information on the specificity of conducting research works on urban agriculture.

As shown by the results of the conducted research, Havana is a unique example of a city where urban agriculture is an integral and important component of the urban substance. Its development, which dates back to the early 1990s and was initially based on a grassroots initiative of the residents, was successively taken over by the government. When faced by the economic crisis during the "Special Period", the Communist country was unable to meet the nutritional needs of its urban population. Therefore, one of the tools for solving the issue of food security was local intra-urban production. The Cuban government encourages cultivation of edible plants on private and state-owned land while ensuring the necessary production resources and trainings within the scope of agrotechnology. By involving the residents of Havana (and other Cuban cities) in the production process, the authorities secured their place among the key actors of the urban food system. Central planning shaped the development of urban agriculture in Cuba and led to its high degree of institutionalisation. The state authorities supervise both where urban gardens are established and how they operate. However, despite the strong institutionalisation and governmental control, urban agriculture in Havana can be considered, to some degree, commercialised. Garden products are predominantly intended for sale and supply the local food market. It is the combination of far-reaching institutionalisation and state control with grassroots movement by residents and almost free access to the food products that determine the distinct nature of urban agriculture in Havana.

Distribution of urban agriculture in the Cuban capital is strongly associated with the spatial development processes that shape it. Both colonial and modernist planning is visible in modern-day urban space of Havana and has indirectly affected where the gardens are located today. Agriculture is distributed unevenly within the city space. It is mainly concentrated in districts with lower building density, in particular those described as the Second Republican Centre near the Revolution Square. Since the 1990s, urban gardens have been filling the gaps left between apartment buildings by modernist planners. Establishment of the Department of Urban Agriculture in 1994 made it easier for residents to obtain the right to use land for food production purposes. From that point, urban agriculture started filling the openings in previously developed urban tissue. Havana, which has neither been destroyed nor thoroughly rebuilt since the city was founded by the Spanish, can be considered a palimpsest, developed as a result of layers from individual periods of its existence being imposed on one another. The network of urban agriculture, whose genesis is secondary in the Cuban capital city, is yet another layer of that palimpsest. Urban gardens are integrated into the urban space, contributing to Havana's multifunctional landscape. However, the specific "gap filling" by agriculture can be viewed in both spatial terms, as making use of abandoned or undeveloped parcels and in functional terms. Urban gardens play an important role in supplying food to the residents of Havana, filling the gaps in the form of food deserts, that is, areas where residents encounter physical and economic barriers in access to food (Thomaier et al. 2015). The numerous facilities in the *Miramar* district are an example of this. Therefore, urban agriculture takes on the supply function, reducing the pressure on rural agriculture in this regard, supported by import from countries of the Eastern bloc. Moreover, it also complements the city's functional structure due to the social role it fulfils.

The results of the aforementioned research also indicate hazards associated with development of urban agriculture when the resources are limited. The dominant food production method in Havana's gardens was organoponics. Its application allows for the use of areas which – due to soil pollution – could not be used for agricultural purposes. Nevertheless, the practice of building the borders of raised plant beds (*canteros*) using asbestos boards and pipes, which is common in the analysed gardens, poses a hazard to life and health of the food producers and consumers. The ubiquity of this harmful material in urban gardens of Havana, which are subject to state control, demonstrates the lack of awareness within the scope of organic production methods on the part of both residents and governmental experts, showing certain ineptitude of a centrally controlled system of urban agriculture in Cuba.

Apart from cognitive results, this chapter also provides important conclusions concerning the research methods applied. The conducted research proves the effectiveness of visual and manual interpretation of satellite and aerial images in Google Earth Pro. It turned out to be an adequate method of analysis of highly heterogeneous vegetation typical of urban agriculture and can be successfully used in similar studies in other cities of the world. Apart from the proven effectiveness of analysis of satellite and aerial images, the research conducted in Havana shows the importance of field observations and the conducted semi-structured interviews. They provided information which would have been difficult or impossible to obtain otherwise than during a direct conversation or observations. The respondents' answers pertained to previously posed research questions and cast light on some issues which had not been considered originally. Therefore, fieldwork should be deemed essential in research on urban agriculture, in particular that which emphasises the human factor.

Notes

1 To facilitate further considerations, this area will hereinafter be referred to as the "city", although its borders do not exactly overlap with the administrative borders designated by Cuban authorities. These borders considerably diverge from the actual concise urban tissue and also include the nearby typical rural areas (peri-urban agriculture and rural agriculture).

2 Taking notes during interviews and making observations in the traditional form (pen and paper) were intentionally avoided so as to prevent situations where the respondent felt officially "interrogated" and the researcher attracted unnecessary attention.

3 As of Monday, 29 May 2023, the current population of Cuba is 11,306,609 (Worldometers 2023, retrieved on 29 May 2023), and the current population of Havana is 2,149,000 (Macrotrends 2023, retrieved on 29 May 2023).

4 The other one was Puerto Rico, taken over by the United States in 1898.

5 The Capitol (Sp. *El Capitolio Nacional de La Habana*) is a building constructed in the 1920s, which was the seat of the Congress until the Cuban Revolution and subsequently became the seat of the Academy of Sciences (Sp. *Academia de Ciencias*) and the Ministry of Science, Technology and the Natural Environment (Sp. *Ministerio de Ciencia, Tecnología y Medio Ambiente*). The building is currently undergoing renovation and is accessible to tourists.

6 This sub-chapter as well as sub-chapters 3.3–3.6 use fragments of an original paper: Górna, A., & Górny, K. (2020). Urban agriculture in Havana–evidence from empirical research. *Miscellanea Geographica*, *24*(2), 85–93.

7 Representatives of the Poder Popular movement, which is an example of participatory democracy, are selected as part of grassroots electoral districts and people's councils from among candidates from each of 168 Cuban municipalities (Cuba Debate 2021, retrieved on 23 January 2021).

8 In the case of *organopónicos de alto rendimiento*, the value is provided for the year 1994.

9 The area of the research zone makes up 10.5% of the entire city, that is, the La Habana province in administrative terms.

10 In 2017, the company assisted in the reconstruction of residents' houses in provinces affected by Hurricane Irma. Asbestos was delivered, among others, to Villa Clara, Ciego de Ávila, Camagüey, Las Tunas and Holguín (Granma Cuba 2017, retrieved on 8 August 2020).

11 Mildew or powdery mildew is a so-called ectoparasite – a fungus-causing disease on the surface of plants, especially vegetables.

References

Alonso González, P. (2016). The organization of commemorative space in postcolonial Cuba: From Civic Square to Square of the Revolution. *Organization*, *23*(1), 47–70, https://doi.org/10.1177/1350508415605100

Altieri, M. A., Companioni, N., Cañizares, K., Murphy, C., Rosset, P., Bourque, M., & Nicholls, C. I. (1999). The greening of the "barrios": Urban agriculture for food security in Cuba. *Agriculture and Human Values*, *16*(2), 131–140, https://doi.org/10.1023/a:1007545304561

Buchmann, C. (2009). Cuban home gardens and their role in social–ecological resilience. *Human Ecology*, 37(6), 705–721, https://doi.org/10.1007/s10745-009-9283-9

Castañeda Abad, W., Herrera Sorzano, A., González Sousa, R., & San Marful Orbis, E. (2017). Población y organoponía como estrategia de desarrollo local [Population and organoponics as a local development strategy]. *Revista Novedades en Población*, *13*(25), 43–55.

Chaplowe, S. G. (1998). Havana's popular gardens: sustainable prospects for urban agriculture. *Environmentalist*, 18(1), 47–57, https://doi.org/10.1023/a:1006582201985

Colantonio, A., & Potter, R. B. (2006). City profile. Havana. *Cities*, *23*(1), 63–78, https://doi.org/10.1016/j.cities.2005.10.001

Coyula, M. (2002). City, tourism and preservation. The Old Havana Way. *ReVista: Harvard Review of Latin America*, 66–69.

Coyula, M., & Hamberg, J. (2003). Urban slums reports: The case of Havana, Cuba. In *The challenge of slums: Case studies for the global report on human settlements*, eds. University College London, London Development Planning Unit, United Nations Human Settlements Programme, London: Authors, 1–40.

Cruz Hernández, M. C., & Medina, R. S. (2001) Agricultura y Ciudad: Una Clave para la Sustentabilidad, Fundación de la Naturaleza y el Hombre [Agriculture and the city: A key to sustainability, nature and man foundation], Havana.

Cuba Debate. (2021). *Poder popular*, http://www.cubadebate.cu/etiqueta/poder-popular/, accessed 23.01.2021.

Currie, L. P. A. (2012). From colonial port to post-revolution: Urban planning for 21st century Havana. *Consilience*, *8*, 50–69.

Czerny, M. (2014). *Stare i nowe w przestrzeni miast Ameryki Łacińskiej. Aktorzy i kontestatorzy zmian* [Old and new in the space of Latin American cities. Actors and contestants of change]. Wydawnictwa Uniwersytetu Warszawskiego.

Del Vas Mingo, M. M. (1985). Las Ordenanzas de 1573, sus antecedentes y consecuencias [The Ordinances of 1573, their background and consequences]. *Quinto centenario, 8*, 83–101.

Díaz, J. P., & Harris, P. (2005). Urban agriculture in Havana: Opportunities for the future. In *Continuous productive urban landscapes*, eds. A. Viljoen, & J. Howe. Routledge, 135–145, https://doi.org/10.4324/9780080454528-29

Edge, K., Scarpaci, J., & Woofter, H. (2006). Mapping and designing Havana: Republican, socialist and global spaces. *Cities, 23*(2), 85–98, https://doi.org/10.1016/j.cities.2005.12.008

Epelman, M. (1993). The export of hazards to the third world: The case of asbestos in Latin America. *NEW SOLUTIONS: A Journal of Environmental and Occupational Health Policy, 2*(4), 48–56, https://doi.org/10.2190/ns2.4.j

Giradet, H. (2012). Urban agriculture and sustainable urban development. In *Continuous productive urban landscapes*, eds. A. Viljoen, & J. Howe. Routledge, 51–58, https://doi.org/10.4324/9780080454528-15

Granma Cuba. (2017). *Redobla Fibrocemento Santiago apoyo a provincias afectadas por Irma* [Fibrocemento Santiago redoubles support to provinces affected by Irma], http://www.granma.cu/cuba/2017-11-05/redobla-fibrocemento-santiago-apoyo-a-provincias-afectadas-por-irma-05-11-2017-21-11-18, accessed 08.08.2020.

Górna, A., & Górny, K. (2020). Urban agriculture in Havana–evidence from empirical research. *Miscellanea Geographica, 24*(2), 85–93, https://doi.org/10.2478/mgrsd-2020-0012

Hernández, J. C. P. (2011). A vision for the future of Havana. *Journal of Biourbanism, I*(1), 93–102.

Herrera Sorzano, A. (2009). Impacto de la agricultura urbana en Cuba [Impact of urban agricultura in Cuba]. *Novedades en Población, 5*(9), 1–14.

Herrera Sorzano, A. H. (2015). La soberanía alimentaria desde la agricultura urbana: un reto para el desarrollo de la producción de alimentos en cuba [Food sovereingnty from urban agriculture: A challenge to the development of food production in cuba]. *Revista GeoNordeste, 1*, 150–172.

Hoyt, H. (1939). *The structure and growth of residential neighborhoods in American cities*. US Government Printing Office.

Jiménez, G., Pérez, M. D. L. A., Nega, H., & Hano, O. (1995). Pesquisaje de lesiones precancerosas y cancerosas del colon en los trabajadores del asbesto [Screening for precancerous and cancerous lesions of the colon in asbestos workers]. *Revista Cubana de Higiene y Epidemiología, 33*(1), 7–8.

Koont, S. (2007). Urban agriculture in Cuba: Of, by, and for the Barrio. *Nature, Society, and Thought, 20*(3/4), 311–325.

Koont, S. (2008). A Cuban success story: Urban agriculture. *Review of Radical Political Economics, 40*(3), 285–291, https://doi.org/10.1177/0486613408320016

Koont, S. (2009). The urban agriculture of Havana. *Monthly Review, 60*(1), 63–72, https://doi.org/10.14452/mr-060-08-2009-01_5

Le Corbusier. (1933). La Charte d'Athènes, édition française 1971. *Paris, Points Seuil*, 190p.

Macrotrends. (2023). *Havana population*, https://www.macrotrends.net/cities/20870/havana/population, accessed 29.05.2023.

Melendo, J. M. A., & Verdejo, J. R. J. (2008). 578: Spanish-American urbanism based on the laws of the Indies: A comparative solar access study of eight cities. In *PLEA 2008 – 25th conference on passive and low energy architecture*, Dublin.

Murphy, C. (1999). *Cultivating Havana: Urban agriculture and food security in the years of crisis*. Food First Institute for Food and Development Policy.

Nelson, E., Scott, S., Cukier, J., & Galán, Á. L. (2009). Institutionalizing agroecology: Successes and challenges in Cuba. *Agriculture and Human Values*, *26*(3), 233–243, https://doi.org/10.1007/s10460-008-9156-7

Novo, M. G., & Murphy, C. (2000). Urban agriculture in the city of Havana: A popular response to a crisis. In *Growing cities, growing food: Urban agriculture on the policy agenda. A reader on urban agriculture*, eds. N. Bakker, M. Dubbeling, S. Guendel, U. Sabel-Koschella, & H. De Zeeuw. Deutsche Stiftung fur Internationale Entwicklung (DSE), Zentralstelle fur Ernahrung und Landwirtschaft. 329–346, https://ruaf.org/document/growing-cities-growing-food/

Plonska, O. (2017). *This garden is who I am. How urban gardeners in Cuba experience their relationship with the state, their gardens and moments of freedom*. VU University of Amsterdam.

Ponce Herrero, G. (2007a). Crisis, posmodernidad y planificación estratégica en La Habana [Crisis, postmodernity and strategic planning in Havana]. *Anales de geografía de la Universidad Complutense*, *27*(2), Universidad Complutense de Madrid, 135–150.

Ponce Herrero, G. (2007b). La ciudad moderna en La Habana [The modern city in Havana]. *Investigaciones Geográficas*, *44*, 129–146, https://doi.org/10.14198/ingeo2007.44.07

Potter, R., & Dann, G. (2000). Tourism, post-modernity and the Caribbean urban imperative. In *The urban Caribbean in an era of global change* , ed. R. Potter, Ashgate: Aldershot and Burlington, 77–102.

Premat, A. (2003). Small-scale urban agriculture in Havana and the reproduction of the "new man" in contemporary Cuba. *European Review of Latin American and Caribbean Studies*, 85–99, https://doi.org/10.18352/erlacs.9695

Premat, A. (2005). Moving between the plan and the ground: Shifting perspectives on urban agriculture in Havana, Cuba. In *Agropolis*, ed. L. J. A. Mougeot, 171–204, https://idrc-crdi.ca/en/book/agropolis-social-political-and-environmental-dimensions-urban-agriculture

Premat, A. (2009). State power, private plots and the greening of Havana's urban agriculture movement. *City & Society*, *21*(1), 28–57, https://doi.org/10.1111/j.1548-744x.2009.01014.x

Premat, A. (2012). *Sowing change the making of Havana's urban agriculture*. Vanderbilt University Press, https://doi.org/10.2307/j.ctv16759bh.7

Rodríguez Castellón, S. (2003). La agricultura urbana y la producción de alimentos: La experiencia de Cuba [Urban agriculture and food production: The Cuban experience]. *Cuba Siglo XXI*, *30*, 77–101.

Rojas, J. A. (1977). Teoría urbanística en la colonización española de América: Las Ordenanzas de Nueva Población [Urban theory in the Spanish colonization of America: The New Population Ordinances]. *Ciudad y Territorio. Ciencia Urbana*, *31*, 1–102, https://recyt.fecyt.es/index.php/CyTET/article/view/81120

Rosset, P., & Benjamin, M. (1994). *The greening of the revolution: Cuba's experiment with organic agriculture*. Melbourne: Ocean Press.

Sánchez Bella, I. (1992). Las Ordenanzas de Felipe II sobre nuevos descubrimientos (1573): consolidación de la política de penetración pacífica. In *De conquistadores: Realidad, justificación, representación*, Vervuert Verlagsgesellschaft, 82–96, https://doi.org/10.31819/9783964566775-006

Scarpaci, J. L. (2000). Reshaping Habana Vieja: Revitalization, historic preservation, and restructuring in the socialist city. *Urban Geography*, *21*(8), 724–744, https://doi.org/10.2747/0272-3638.21.8.724

Segre, R., & Baroni, S. (1998). Cuba y La Habana. Historia, población y territorio [Cuba and Havana. History, population and territory]. *Ciudad y Territorio Estudios Territoriales*, 351–379.

Singh, S., Singh, D. R., Velmurugan, A., Jaisankar, I., & Swarnam, T. P. (2008). Coping with climatic uncertainties through improved production technologies in tropical island conditions. In *Biodiversity and climate change adaptation in tropical islands*, eds. C. Sivaperuman, A. Velmurugan. A. K. Singh, & I. Jaisankar, Academic Press, 623–666, https://doi.org/10.1016/b978-0-12-813064-3.00023-5

Tablada, A., De Troyer, F., Blocken, B., Carmeliet, J., & Verschure, H. (2009). On natural ventilation and thermal comfort in compact urban environments–the Old Havana case. *Building and Environment*, *44*(9), 1943–1958, https://doi.org/10.1016/j.buildenv.2009.01.008

Thomaier, S., Specht, K., Henckel, D., Dierich, A., Siebert, R., Freisinger, U. B., & Sawicka, M. (2015). Farming in and on urban buildings: Present practice and specific novelties of Zero-Acreage Farming (ZFarming). *Renewable Agriculture and Food Systems*, *30*(1), 43–54, https://doi.org/10.1017/s1742170514000143

Tweedale, G., & McCulloch, J. (2004). Chrysophiles versus chrysophobes: The white asbestos controversy, 1950s–2004. *Isis*, *95*(2), 239–259, https://doi.org/10.1086/426196

Viljoen, A., & Bohn, K. (2014). Utilitarian dreams: Food growing in urban landscapes. In *Second nature urban agriculture*. Routledge, 32–39, https://doi.org/10.4324/9780080454528-36

Viljoen, A., & Howe, J. (2005). Cuba: Laboratory for urban agriculture. In *CPULs. Continuous productive urban landscapes: Designing urban agriculture for sustainable cities*, ed. A. Viljoen. Oxford: Elsevier, 146–191, https://doi.org/10.4324/9780080454528-30

Wolf, E. R. (1982). *Europe and the people without history*. University of California Press, https://doi.org/10.2307/493157

Worldometers. (2023). *Cuba population*, https://www.worldometers.info/world-population/cuba-population/, accessed 29.05.2023.

4 Singapore – a case study of urban agriculture in Southeast Asia

The case study analysed in this chapter is Singapore – an East Asian city-state, considered one of the most dynamically developing metropolises in the world. Due to the country's small area as well as the authorities' policy aimed at highly efficient use of the limited spatial resources, the presence of agriculture within its borders may appear to be a peculiar or at least unusual form of land use. What is also important is the fact that urban agriculture is treated as an activity that generates minor economic benefits, as described more broadly in Chapter 1. However, taking into account the share of food Singapore is forced to import from abroad, promotion of local food production should be considered a reasonable approach. According to data published by the Singapore Food Agency (SFA) in 2018, this share amounted to 90% (Singapore Food Agency 2021b); therefore, the city-state's authorities promote urban agriculture, but only in its most modern form – that of lucrative farms using cutting-edge technologies. On the other hand, urban farms and gardens established at the residents' initiative are still present in Singapore. They are not aligned with the government's policy and their continued operation is at risk more than that of other establishments. Features of urban agriculture and its role in the spatial and functional structure are shaped by the opposing interests of the authorities and the residents' needs. However, it should be emphasised that under a semi-authoritarian regime, if those needs do not align with the authorities' current policies, they are not treated as priorities and frequently end up not being met.

Detailed analysis covered 36 urban farms, which were visited during field research carried out at the turn of January and February 2019.[1] The method of visual classification, applied in all case studies discussed in the book, proved less useful in Singapore than it was in the case of Havana and Kigali. In Singapore, where crop cultivation or animal husbandry are typically pursued under a semi-permeable cover or a roof, it was necessary to supplement the research with an online query, which allowed for precise determination of individual farms' locations. The research area designated in Singapore overlapped with the city's administrative borders, which are not identical to the country's borders. The research excluded smaller islands which are uninhabited, fulfil recreational functions, for example, Sentosa, or industrial islands, for example, the artificial island of Jurong. Contrary to Havana and Kigali, developed land in Singapore is not surrounded by agricultural areas but by

DOI: 10.4324/9781003429845-4

the coastline. Urbanisation of this city-state is so advanced that the entire island is classified as an urban island (Motha & Yuen 1999). For this reason, distinguishing between intra-urban farming and peri-urban or rural farming was not necessary in the case of this Asian city-state.

Semi-structured interviews were conducted at 18 out of the 36 visited urban farms. For the remaining ones, research was limited to field observations and collecting photographic documentation. The group of respondents comprised farm managers or owners and their employees. The semi-structured interviews were conducted according to a list of topics prepared in advance. Similar to Havana, thanks to the possibility for new subjects to be introduced by both the respondent and the researcher, comprehensive information on the functioning of individual farms was supplemented with issues pertaining to problems and challenges faced by the actors involved in urban agriculture in Singapore. Additional valuable information was obtained during an interview with the Executive Manager of the Food Supply Resilience Group, an organisational unit of the Agri-Food and Veterinary Authority (AVA) of Singapore. The interview was conducted online and provided valuable insight into the authorities' policy with respect to urban agriculture and opportunities that the institution provides to residents who wish to engage in the process of local food production.

The following sub-chapters will first provide a comprehensive characterisation of Singapore's spatial and functional structure, discuss the measures taken by the authorities of the city-state within the scope of implementing the assumptions of the smart city concept and the institutional and legal framework shaping the directions of development of urban agriculture. The subsequent sub-chapters, on the other hand, contain a detailed discussion of the research results.

4.1   Spatial and functional structure of the city[2]

Singapore is situated on several islands at the southeast coast of the Malay Peninsula, between continental Malaysia on one side (separated from the islands by the Johore Strait) and the Indonesian Riau Archipelago on the other side (separated by the Singapore Strait). The city-state encompasses the Singapore Island with an area of 710 km² and 63 smaller islands (Henderson 2012). Most of them are uninhabited and not exploited economically. The precise area of Singapore is difficult to determine due to continuous anthropogenic expansion of its territory. It results from the shallows and coastal reefs being reclaimed and replaced by artificial islands. There are currently already ten of them. This practice, typical of many dynamically developing Asian countries, gives rise to a number of concerns regarding its impact on the natural environment as well as political implications (Chee et al. 2017). Land reclamation in Singapore can lead to a significant decline in biodiversity and destruction of natural habitats (Lai et al. 2015). According to the official data of the Singapore Department of Statistics collected by the Singapore Land Authority, in 1970, when an administrative decision was made to align the country's borders with the borders of the capital city, its area amounted to 586.4 km². This decision is considered to be the moment of establishment of the city-state. In 2020, the area of the country amounted to 728 km², which means that within a half-century, Singapore had grown by nearly 142 km², that is, 24%. According to the latest data as of December 2022, in turn, the area of Singapore is 734.3 km² (Singapore Department of Statistics 2023).

Singapore is divided into 55 planning areas, organised into five regions – Central, East, North, North-East and West. Moreover, two catchment areas for rainwater collection have been designated on the island. The first one – Central Water Catchment – is located in the central part of the city and contained within the borders of the Central Region. It features four reservoirs: MacRitchie, Upper Seletar, Upper Peirce and Lower Peirce, as well as recreational facilities, such as a zoo or night safari. On the other hand, the second catchment area – Western Water Catchment – is located in the west part of the island in the West Region and comprises mostly the military training area of the Singapore Armed Forces. There are four reservoirs within its limits: Tengeh, Poyan, Murai and Sarimbun. However, the listed regions are not units of administrative division, according to which Singapore is divided into districts whose borders change depending on the results of population censuses. Despite the foregoing, the majority of governmental institutions use regions as areas of reference due to the fact that their borders do not undergo regular changes.

When analysing the contemporary spatial and functional structure of the city, one should note the stages of its development, starting from first pre-colonial settlements in the region through the colonial period initiated with location of the city by the British in 1819, to times of independence and contemporary pro-environmental reforms introduced by the Singaporean authorities.

4.1.1 The pre-colonial period

The island on which Singapore is located today was already marked by Ptolemy in the *Geographia* atlas; however, the first mentions of settlements in the modern-day Singapore area come from Javanese and Chinese written sources dating back to late 14th century. According to those Asian texts, the island, referred to as *Tumasik* or *Tumasek*,[3] was originally inhabited by fishers and pirates and was used as an outpost of the Sumatran Srīvijaya Empire at the coast of the Malay Peninsula (Santhi & Saravanakumar 2020). In the 13th and 14th centuries, the island was attacked multiple times by the Javanese army, and in the second half of the 14th century, it was annexed by the Hindu-Buddhist Majapahit Empire (Turnbull 2009; Abshire 2011). Under its rule, the island lost its prominence and was incorporated first into the Malacca Sultanate, then into the Johor Sultanate (Santhi & Saravanakumar 2020). One of the first meetings between the indigenous inhabitants of Singapore and Europeans is dated at the beginning of the 17th century, when Portuguese explorers burnt down the local trading port. The European attack led to nearly 200 years of stagnation of Tumasik, which from that point was inhabited mainly by Malaysian officials of the Johor Sultanate and Chinese farmers (ibidem).

4.1.2   British colonialism

The turning point in Singapore's development and, at the same time, the factor which led to the establishment of permanent buildings was the arrival of Sir Thomas Stamford Raffles on the island in 1819. Raffles (1781–1826) was the deputy governor of the British enclave of Bencoolen on the west coast of Sumatra and a representative of the British East India Company (Turnbull 2009; Abshire 2011). At the Company's order, Raffles was supposed to select a place suitable for establishment of a European trading port. Singapore's location was exceptionally advantageous from the strategic and communication point of view. The island is located at a junction point of Asia, at the intersection of main trading routes through the Indian Ocean, connecting Europe, Africa and the Middle East with China, Japan and Australia (Eng 1986; Turnbull 2009). This excellent location prevailed over the lack of natural resources and small area of the island and has been the main driver of Singapore's development for centuries. Establishment of a European trading port contributed to a demographic and economic boom in the city, which was already purchased by the British East India Company in 1824. In 1867, seven years before the Company's dissolution, Singapore became the property of the British government, and thus a colony of the British empire. The city soon became the largest British commercial and political hub in Malaya (Henderson 2012).

Thomas Stamford Raffles is also associated with the beginnings of European spatial planning in Singapore. In 1822, the deputy governor initiated the spatial development plan called the "Raffles Town Plan" or "Jackson Plan", named after its author – Lieutenant Philip Jackson (Chew 2009). The document, published in 1828, was meant to ensure orderly spatial development of the city, which was experiencing intensive demographic growth. Its main assumptions included a regular, orthogonal street grid and spatial segregation of different ethnic groups – Europeans, Chinese, Arabs, Malays and Indians. The road layout and the Commercial Square set out at the time (whose name was eventually changed to the Raffles Place), proposed in the Jackson Plan, are to this day visible in the urban tissue of Singapore's oldest districts (ibidem). Moreover, the contemporary Little India, Chinatown and Arab Street are also the legacy of the first decisions concerning spatial ethnic segregation in the city. During his first months in office in Singapore, Raffles allocated an area of approximately 19 ha to a botanical garden, which to this day remains one of the key concise green areas in the central districts of the city (Leitmann 2000), as well as the only Singaporean site included on the UNESCO World Heritage List (it was added to this prestigious list in 2015).

One of the key thoroughfares of modern-day Singapore – Orchard Road – is important in the context of this chapter. An exclusive commercial zone has now developed along this road, which is compared to the Oxford Street in London, with renowned stores, boutiques, restaurants, hotels and cafes. However, in the past, Orchard Road was strictly associated with the development of agriculture in British Singapore (Leitch Lepoer 1989). The street was set out in the 1830s, which coincided with a period of dynamic increase in demand for spices. Nutmeg and pepper plantations were concentrated along this road, as were fruit orchards and vegetable gardens. The road is most likely named after one of the plantation owners, whose surname was Orchard. However, it should be noted that the English word "orchard" suggests a realistically motivated origin of the name (Cornelius 1999). Plantations and farms along the Orchard Road already started declining in the second half of the 19th century as a result of lower demand for spices, as well as an intensive spatial development of Singapore. The land previously used for agricultural purposes was replaced by residential buildings, while farms were gradually removed to areas more and more remote from the historical city centre.

In the early 20th century, Singapore experienced a dynamic increase in population, which resulted in serious overpopulation. Due to the lack of an updated development plan that would allow the British to control spatial development, marginal districts began to form in the city (Leitmann 2000). Only in 1927 was the Singapore Improvement Trust established. Its objectives included construction and widening of roads, which were too narrow for the increasingly popular car transport, as well as demolition of buildings with low sanitary standards, that is, slum clearance (Cheng 1995; Leitmann 2000; Neo 2022). In spite of constructing more than 23,000 residential buildings, the trust was still unable to meet the needs of the continuously growing population (Leitmann 2000). The situation in the city was significantly deteriorated by the wartime destruction of infrastructure caused by military operations during World War II. In 1942, the city was attacked by the Japanese army, which occupied Singapore after a long bombing, causing the defending British forces to suffer the greatest military defeat in the history of the empire (with approximately 80,000 British soldiers taken prisoner). The Japanese occupation in 1942–1945 was marked by heavy bombings by the Allied forces, resulting in a large portion of the city being razed to the ground (Davies 2017). However, the years following the war proved to be the beginning of a period of prosperity for Singapore. The development of rubber trade and the city's role as a base for the Allied navy during the Korean War (1950–1953), fought several thousand kilometres to the north-east of the city, contributed to the gradual enrichment of the British colony (Santhi, Saravanakumar 2020). In 1951, work commenced on the development of a new planning document that would bring order to the city's dynamic development. The Statutory Master Plan, completed in 1955 and approved in 1958, regulated the type and intensity of development by specifying the terrain allocation and the maximum building density on individual parcels. Implementation of the document's provisions was managed centrally by the newly established Planning Department, which had been granted the authority to control land development throughout the island (Chew 2009). In 1959, the Planning Ordinance was adopted, which led to the establishment of a central body supervising spatial development in Singapore (Chew 2009; Yuen 2009).

4.1.3 *Independence and green reforms*

After Singapore gained independence from Great Britain and Malaysia on 9 August 1965, it was quickly put on the track of dynamic economic growth. Since the 1960s, the urban landscape has undergone major transformations while the city-state was being rapidly transformed into a modern metropolis (Eng 1986). In terms of spatial planning, the goal of the new government, led by Lee Kuan Yew, the prime minister in 1965–1990, was primarily to provide the residents with healthy living conditions, make more efficient use of space and develop commercial, industrial, residential and recreational areas (Kong, Yeoh 1994). In 1966, the Urban Renewal Department of the Housing and Development Board was established. In 1974, it became an independent governmental administrative body. The first years of its functioning were focused primarily on dismantlement of damaged buildings, slum clearance, resettlement of people from central districts and construction of new housing estates planned so as to ensure the most efficient and intensive development of the limited space (ibidem). The dynamic development of the city also came at the cost of agricultural areas. Farms were being moved to more and more peripheral locations and the government intentionally gave up on local production in favour of importing food products. By 1988, the agricultural land in Singapore had been built over and its areas dropped to approximately 3% of the city's area (Leitch Lepoer 1989). Another crucial decision of the authorities was introduction of the ban on pig breeding in the city in 1984, justified by the unpleasant smell and pollution caused by the animals.

Issues associated with protecting Singapore's natural environment have been taken into account in planning documents and practices since the beginning of its existence as an independent state. This was a necessary measure in the context of the legacy of the colonial period, after which the colony was left practically without any fragments of concise, natural vegetation of the equatorial forest that used to grow there (Kong & Yeoh 1996). The process of mangrove deforestation near Kranji in the north-west part of Singapore and Tuas in the west part as well as logging of the freshwater swamp forest in the lowlands regularly flooded with fresh water, among others, near Saletar in the north-east part of the city, was initiated in the first half of the 19th century by the city's founder, Stamford Raffles. He strived to take over the vast swathes of land at the edge of the Malay Peninsula and use them as plantations (of rubber trees, among others). In subsequent decades, those plantations were replaced by dynamic urbanisation (Kong & Yeoh 1996).

Already in 1963, the Singaporean authorities initiated a period of pro-environmental measures with a tree planting campaign (Tan 2006; Tan & Neo 2009; Tan et al. 2013). On the other hand, after obtaining independence in 1965, selected assumptions of the Garden City concept of E. Howard (1902) began to be incorporated into Singapore's planning documents, which ultimately led to the establishment of the Garden City Action Committee in 1973, overseeing the progress in their implementation (Savage & Kong 1993; Leitmann 2000; Tan 2006; Tan et al. 2013). However, it should be emphasised that activities involving introducing greenery to developed spaces, taken up in the 1960s, were extremely limited. As a result of spontaneous campaigns, only small parcels, which were the last fragments of land unsuitable for intensive development, were typically transformed into city parks.

The 1970s were characterised by more integrated and organised measures coordinated by the authorities (Henderson 2013). In 1972, the authorities of Singapore were among the first ones in the world to establish the Ministry of Environment,[4] proving the importance of issues of environmental protection in their policy (Savage & Kong 1993; Leitmann 2000). Greenery was introduced into the city over subsequent decades and Singapore has now become one of the model examples of a large metropolis organised in the spirit of a Garden City, where vegetation in the form of city parks of various sizes, connected with vast ribbons of urban greenery, is the foundation of the urban ecosystem (Tan 2006; Henderson 2012). It is thanks to those linear connectors, which form wildlife corridors, that the network of urban greenery in Singapore is described as a park connector network. However, the city-state's vegetation consists practically exclusively of recreated greenery. Primal vegetation only makes up 0.28% of all greenery (Tan et al. 2013). Its high availability, achieved thanks to the green connector network, is particularly important in the context of the residents' quality of life (Tan 2006). A document entitled "The Concept Plan", published by the Urban Redevelopment Authority in 1991, assumed that in view of Singapore achieving a high level of social and economic development, the next priority should be increasing the density of the network of green recreational areas and water arteries (Singapore Government Agency 1991). A series of measures to this end has been described as the "Green Plan" or "Green and Blue Plan", in reference to development of green areas, water reservoirs and watercourses in the city (Rowe & Hee 2019). The aforementioned plan replaced the previous environmental planning strategy in Singapore. Since its introduction, domination of a holistic approach (instead of previous typical ad-hoc measures) has been observed, integrating environmental education activities within measures in terms of green and blue infrastructure and eco-friendly technologies (Leitmann 2000).

In connection with the widespread reforms concerning creation of the urban ecosystem structure, included among others in "The Singapore Green Plan 2012", the city is considered one of the pioneers of the new trend of biophilic urbanism (Tan et al. 2013; Newman 2014). It is based on the assumption that the natural environment should be at the centre of spatial planning. Moreover, thanks to introducing greenery on building rooftops and walls, Singapore is also referred to as a Vertical Garden City.

4.1.4 Singapore's functional centres

The spatial and functional structure of contemporary cities was shaped, to a large extent, by the development of transport infrastructure (Anas et al. 1998). As a result of the industrial revolution, the directions of spatial expansion of many cities were determined by the routes of railway lines. On the other hand, during the period of Fordism and post-Fordism, this role was taken over by roads intended for car transport and the public transport network. According to C. Zhong et al. (2014), spatial development of Singapore was closely connected with the development of transport and communication network planning has already been used as a tool for shaping the city's spatial structure since the 1970s. Residential districts with high building density were purposely located near high-capacity roads and in close proximity to industrial plants and other places of employment (ibidem). An important moment for the spatial development of Singapore was the year 1987, when the first system of public rail transport – Mass Rapid Transit (MRT) – was opened. The city currently has 122 functioning underground and aboveground railway stations, supported by a dense network of bus stops. Since half of the island's residents use public transport on a daily basis (Cheong & Toh 2010), this network serves as a frame connecting all regions of Singapore (Zhong et al. 2014). The development of public transport was used to implement the urbanistic assumptions prepared in the 1990s. They included decentralisation of Singapore's urban tissue, planned on a top-down basis. According to the assumption, the city centre was supposed to be surrounded by four regional hubs located in the north, west, north-east and east, as well as several sub-centres on the edges of the island (ibidem). It was assumed that the location of new commercial centres within the island would reduce the distances between places of work and recreation and housing estates, while at the same time decreasing the stress on the business district (Singapore Government Agency 1991).

When identifying the functional centres of Singapore, C. Zhong et al. (2013) propose an overview of the city's morphological features combined with the residents' spatial behavioural tendencies. In their opinion, functional centres are areas characterised by intensity and diversity of urban activities. In order to determine the degree of centrality of a given space, the authors propose combining two measures – density and diversity of the population's activity. They list a number of functional centres characterised by different levels of centrality (ibidem).

The first one, referred to as Downtown, is located in the southern part of the island and covers areas such as Outram and Downtown Core, located on the Marina Bay, as well as Orchard, situated along one of the city's main thoroughfares – Orchard Road. Downtown is also the oldest part of the city, since it is included on development plans drawn up still during the colonial times. In the early 1990s, fragments of Downtown located on the Singapore River underwent rejuvenation (also referred to as urban reclamation), according to the concept of multifunctionality of the waterfront (Chang & Huang 2011). It involved, first and foremost, introduction of new functions, maximisation of economic usefulness of the space, as well as improvement of the space's availability to its users (ibidem).

Four functional sub-centres designated by Zhong et al. (2013) are the following areas: Jurong in the East Region, Tampines in the West Region, Woodlands in the North Region and Seletar in the North-East Region. All of the aforementioned sub-centres were established in accordance with the spatial planning assumptions implemented by the authorities since the 1990s, which proves compatibility of spatial planning and planning of the transport network (ibidem), as well as the effectiveness of the authorities in creating urban reality. Indeed, it was the authorities that assumed decentralisation of Singapore and planned the placement of the new centres, which to a large extent has been successfully achieved, as demonstrated by the referenced authors. However, they also note that the level of centrality of Yishun (located in the northern part of the island), which developed on a bottom-up basis, is higher than that of the Woodlands centre, planned on a top-down basis, which may (but does not need to) demonstrate that the residents themselves have a major impact on the urban tissue. Some fragments of Singapore are also characterised by extremely low level of centrality. These are primarily the catchment areas in the centre (Central Water Catchment) and in the west (Western Water Catchment) of the island. Another such fragment is an area of importance from the point of view of this book – Kranji Countryside, an industrial and agricultural area located in the north-west of the island. It is poorly connected with the remaining districts of Singapore and, at the same time, characterised by a low level of diversity of urban activity. It is within that area that the highest share of urban farms has been observed.

On the basis of the referenced studies, the spatial and functional structure of Singapore can be considered polycentric, while the government's measures in terms of decentralisation of the city can be deemed effective. Although some of the sub-centres planned in the early 1990s have not been fully developed to this day, synchronisation of spatial planning and planning of the transport network has yielded the desired results. Development of the public transport system, in particular the subway line, was an important factor that shaped not only how the space is used, but also the diversity and spatial concentration of the residents' activity within the city limits.

The planning document, which is currently being implemented for Singapore, is the Master Plan published in 2019. Its assumptions fit in with the concept of sustainable development, with particular emphasis on measures within the scope of mitigating the impact of global warming, as well as with the concept of resilient and smart cities. Moreover, the document strongly emphasises supporting open and inclusive societies as well as construction of green housing estates.

Singapore today is facing a number of challenges caused by the continuous population growth combined with limited space, which require highly effective, integrated spatial planning (Henderson 2012; Tan et al. 2013). Due to the growing housing needs, one of the most frequently applied solutions is increasing building density. In spite of measures aimed at preserving greenery within the city space, in 1965–2000, the area of developed land in Singapore doubled at the expense of not only forests, but also agricultural areas (Roth & Chow 2012). Due to extremely high competition for land, activities that are not economically efficient enough are unable to remain within the city space. Thus, land development affects primarily the subject of this book's analysis – urban agriculture. Due to the fact that Singapore is forced to import nearly all of its food (Tey et al. 2009), local food production should be one of the priorities for the current authorities. It allows for a certain level of independence from supplies of food products from abroad, shortening of the supply chain and reduction of transport, and thus also of the carbon footprint. Nevertheless, due to the insufficient land resources and high level of competition on the part of other, more lucrative forms of land use, many urban gardens and farms are incapable of remaining on the market. Urban agriculture in Singapore is therefore in a difficult position, created by the existing institutional and legal framework and high competitiveness on the market. *Summa summarum*, both factors necessitate reorganisation of the land use structure in favour of urban buildings of higher priority.

4.2   Singapore – a smart city

The term "smart city" was first used in the 1990s (Gibson et al. 1992). It was also then that it became inseparable from the implementation of new Information and Communications Technologies (ICTs) in the urban infrastructure (Albino et al. 2015). However, this technocratic approach is nowadays criticised and modern technological solutions are no longer treated as a central element of the smart city concept, which has become more defined over time (Albino et al. 2015; Monfaredzadeh & Berardi 2015; Maye 2019).

At present, as a consequence of criticism of the technocratic approach, cities that are considered smart are those that use cutting-edge technologies in pursuit of sustainable development, improving the residents' living conditions, improving the quality of the natural environment, as well as generating better prospects for business entities (Monfaredzadeh & Berardi 2015). Significant emphasis is also placed on strengthening social bonds and investments in human and social capital (Albino et al. 2015). In literature, there are six features (sometimes called components or dimensions) typical of smart cities. They were proposed by R. Giffinger et al. (2007) as part of a project carried out by the Regional Science Center at the Vienna University of Technology. Those dimensions are smart economy, smart mobility, smart governance, smart environment, smart living and smart people. In 2014, B. Cohen, in order to operationalise the concept, developed the so-called Smart City Wheel and proposed a number of indicators for each area of the smart city to enable measurement of "city smartness". The study by B. Cohen is the basis for a lot of research concerning smart cities, carried out in various regions of the world (see, e.g., Govada et al. 2017; Shah et al. 2017; Qonita & Giyarsih 2023).

Urban agriculture as an activity enabling sustainable use of urban resources, increasing the share of greenery in the city space, improving the living conditions of residents, as well as contributing to the shortening of the value and supply chains, generates tangible economic benefits – and thus fits in with the assumptions of the smart city concept. Moreover, D. Maye (2019) combines the concept of a smart city and urban food systems (which include urban agriculture), proposing the term "smart food city". Many authors indicate the importance of use of modern technologies and innovative methods of food production in cities as an element of implementing the smart concept (Dos Santos 2016; Kumar & Dahiya 2017; Shamshiri 2018). These technologies enable monitoring and strict control over the crops, which, in turn, helps to increase their efficiency while simultaneously reducing energy consumption losses. However, it seems than an important question to ask is whether other, more traditional forms of urban agriculture, such as community gardens or small farms specialising in soil cultivation, can also be considered consistent with the discussed concept. They frequently result from grassroots initiatives of the residents, and by increasing the share of greenery, they improve the quality of the natural environment and thus of life in the city. This is the approach taken by D. Maye (2019). The author emphasises that the idea of a smart food city assumes primarily the existence of communal and civic forms of innovation within the scope of land use in line with the traditions of the urban food system. Therefore, one can conclude that if urban agriculture (even in its traditional form) is an expression of a grassroots civic initiative of the residents, it aligns with the concept's assumptions.

For years, Singapore has been presented in international rankings as a model smart city and one of the pioneers in this regard on a global scale. In the IMD Smart City Index ranking, developed by IMD World Competitiveness Center, this Asian city claimed the first place among smart cities for three consecutive years – 2019, 2020 and 2021 (IMD 2019, 2020, 2021). In the 2023 ranking, on the other hand (no ranking was published in 2022), the city claimed the 7th spot. The Lion City owes its high position in subsequent rankings not only to the implementation of many modern ICT solutions and fast pace of economic growth, but primarily to the authorities' endeavours to ensure a high quality of life for the residents. Singapore was already called an "intelligent island" at the end of the previous millennium (Mahizhnan 1999, p. 14), when both the level of technological advancement and the scope of technological intervention in people's everyday lives were much smaller than they are nowadays. Since that time, the assumptions of the smart city concept themselves have evolved and the most advanced hubs currently tend to be referred to as "next-generation smart cities", where the focus has shifted from technology to residents and particular emphasis is placed on their participation in managing and creating the city's future. In line with this new idea, in 2014, Singaporean authorities proposed the Smart Nation programme (Smart Nation and Digital Government Office 2023). Although it was to a large extent a continuation of the previous policy, it partially changed the optics of the city-state's development. Since that time, modern technologies have been meant to merely assist in the creation of development perspectives for all residents, while simultaneously supporting them in leading a comfortable and sustainable life (Woo 2017). However, the seemingly civic-oriented but in reality technocratic solutions introduced by the authorities of Singapore are criticised and sometimes seen as anti-democratic measures aimed at the consolidation of power as part of the authoritarian rule of the People's Action Party, while simultaneously reinforcing the "pragmatic" and depoliticised ethos of obedient and disciplined Singaporean society (Ho 2017).

Implementation of measures that align with the smart city concept, in particular in the areas of "smart people" and "smart environment", affects the specificity of urban agriculture in Singapore, as demonstrated by the semi-structured interviews conducted for the purposes of this chapter with representatives of individual urban gardens and the interview with the representative of the Agriculture and Veterinary Authority. However, the current policy of the city's authorities in terms of land use and the competition for space associated with it, as discussed below, turn out to be much more important for the presence of agriculture within the city's spatial structure.

4.3 Institutional and legal framework and the authorities' policy with respect to urban agriculture

When Singapore gained independence in 1965, it was struggling with many issues. The most crucial ones included overpopulation, extremely limited options for spatial development, lack of adequate infrastructure and lack of natural resources, which had to be imported. However, the authorities of this island city-state, using its extremely advantageous location at the intersection of prominent trade routes, have managed to transform the economy of Singapore from one that was previously based on labour-intensive processing of goods in light industry into one focused on the services sector, with emphasis on banking, research, tourism, transport and trade (Shatkin 2014). After the country gained independence from Great Britain and Malaysia, the People's Action Party came to power and has been ruling without interruptions to this day. Initially, the authorities' policy was focused on gaining full control over land and methods of its use, which allowed them to control the development directions of Singapore as a whole very effectively. Already in 1966, the Land Acquisition Act (LAA) significantly facilitated land nationalisation (Shatkin 2014). Currently, all land-related transactions are managed by a national institution, that is, the Singapore Land Authority or Urban Redevelopment Authority (Shatkin 2014; Haila 2015). In 1992, 80% of land belonged to the state (Han 2005). This number currently stands at 90% (Haila 2015). Such extensive competences of the authorities in terms of land management not only generate profits for the state treasury, making it easier for the current ruling party to remain in power, but also allow for highly centralised and thus effective albeit top-bottom spatial planning. This spatial planning model is exported to countries which are facing serious developmental challenges, including mainly other countries of the Global South, such as Rwanda, whose capital city is discussed in the subsequent chapter of this book (Shatkin 2014).

Control over the land resources means that the authorities of Singapore are the main decision-maker in terms of location of agricultural activity within the city limits. In the context of its future, one of the provisions contained in the latest Master Plan published in 2019 is crucial (Government of Singapore 2019). According to it, the following are considered forms of agricultural activity in Singapore: agrotechnology park, aquaculture farm, plant nursery, hydroponics farm and agriculture research/experimental station. Therefore, the document does not take into account traditional forms of agriculture, such as farms practising soil cultivation, which are present within the city. Agriculture in Singapore will therefore be undergoing a number of crucial changes in the near future, not only quantitative or spatial, involving liquidation or relocation of some farms, but also qualitative, leading to changes in the applied production methods. According to research by L. Y. Astee and N. T. Kishnani (2010), the majority of buildings at municipal housing estates in Singapore are adapted to the implementation of rooftop agriculture with the use of soil-free hydroponics, which would significantly increase local food production. Therefore, it is a possible direction of development of urban agriculture on the island, in line with the current policy of the authorities. However, it first requires taking widespread measures in cooperation with state-owned, cooperative and private entities.

Until recently, the main institution in Singapore that business activity associated with agriculture was subject to was the AVA of Singapore. On 1 April 2019, the organisation was divided into two separate entities: the SFA, whose activity is focused on issues related to food production and distribution, and the Animal &

Veterinary Service (AVS), which provides services within the scope of animal care that are unrelated to the food production process. Both institutions provide technical and technological support as well as ensure strategic research and development cooperation. They also have two research units in the Kranji Countryside area. Local farmers can use the services of a dedicated advisor within the scope of implementation and testing of modern technologies, as well as business development and obtaining financial assistance as part of the Agriculture Productivity Fund (APF).

It is true to say that the Singaporean authorities have a number of tools that allow them to control and steer the development of urban agriculture. However, strong centralisation prevents grassroots initiatives and limits the agency of the residents, whose individual needs are typically not taken into account in the strategies pursued by the national institutions. The conducted field research confirms that the institutional and legal structures, although they are far less elaborate than those in Cuba, have a crucial impact on the distribution and features of urban agriculture in Singapore. At this point, it should also be emphasised that Singapore – one of the global leaders within the scope of innovation, the residents' income level as well as social development (12th position in the ranking of countries in terms of HDI in 2021/2022; United Nations Development Programme 2022) – is an authoritarian state. In spite of the global image of a highly developed country, it does not meet the basic norms in terms of civic freedoms, social participation and political rights. As noted by the British historian N. Davies, economic and material criteria should not be the only measure of a country's success (Davies 1998). M. W. Solarz (2014) shares this view. This issue has also been pointed out by authors from the Global South. An Indian Nobel Prize winner, A. Sen, emphasises that development should also be understood as a process of expanding freedoms (Sen 2001). This is echoed by the President of Tanzania J. Nyerere, who wrote that freedom and development depend on one another and the latter cannot exist without the former (Nyerere 2000).

4.4   Distribution of agriculture within the urban space

In the context of this book, the location of agriculture within the urban space is an important information medium. On the one hand, it is the result of the authorities' actions within the existing institutional and legal structures as well as external factors affecting urban agriculture, such as availability of land with specific attributes or the socioeconomic situation. On the other hand, the location itself influences the internal features of agriculture, such as production methods or the organisational form. It entails a number of messages that make up the bigger picture of agriculture's existence within the city.

The characteristics of distribution of urban agriculture in Singapore differ considerably from those in Havana or Kigali. First and foremost, the majority of farms are concentrated in the north-west part of the city (see Figure 4.1). Central districts, in turn, feature a small number of farms due to high building density, high land prices and small supply of space that can be allocated to agricultural activity.

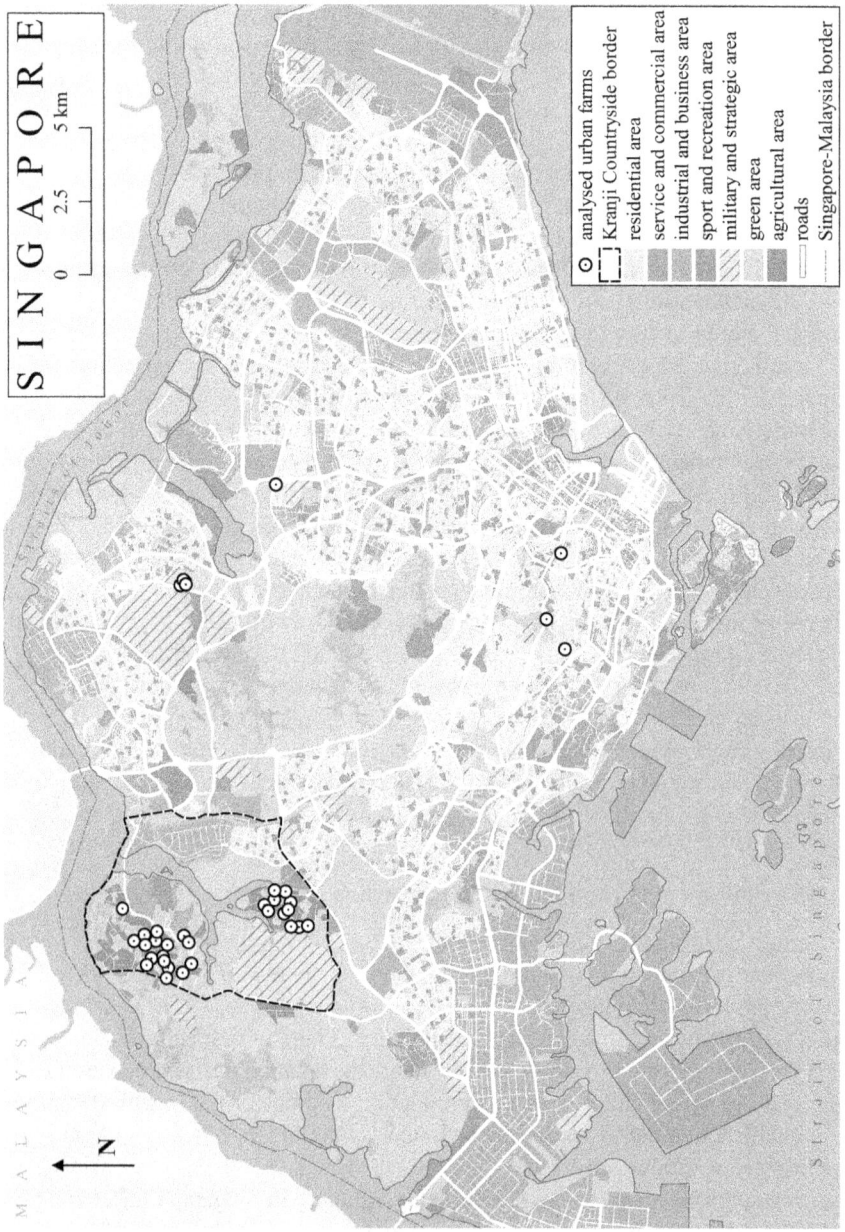

Figure 4.1 Land development and distribution of urban farms in Singapore.

According to the planning division of Singapore, as many as 21 of 36 urban farms covered by detailed field research are in the North Region. In the West Region, there are 11 of them, in the Central Region – only three and merely one in the North-East Region. No farming facility was analysed in the East Region. The 36 urban farms covered by detailed research can be divided into two groups, distinguished based on their location. The first group includes the 28 farms located in the north-west part of the city called Kranji Countryside. This area, which also includes two nature reserves – Sungei Buloh Wetland Reserve and Kranji Marshes – is characterised by a low share of areas allocated to residential development in the overall land area and is meant to serve as a place of recreation, interaction with nature and relaxation away from the city noise for Singaporeans.

However, contrary to what its name suggests, Kranji Countryside can hardly be considered a typical rural location. In spite of the relatively high (for Singapore) number of urban farms within its limits, the majority of them are industrial farms, where the plants grown and animals bred are located in large halls, greenhouses or under permanent cover. Kranji Countryside is Singapore's production base, located at a distance from the city centre. Due to the sparse network of public transport, it is not an easily accessible area, separated from other districts, where the basic mode of transport for people and goods is the car. Urban farms that produce food (of plant and animal origin) as well as those engaged in growing decorative plants and breeding ornamental fish are adjacent to the local military sites and industrial plants, creating a distinct industrial and agricultural landscape.

Figure 4.2 presents forms of land use within the limits of Kranji Countryside. The marked locations are urban farms involved in plant production (divided into areas occupied by crops cultivated directly in the soil, hydroponics and cultivation of decorative plants) as well as animal production (animal husbandry and aquaculture, frog and crocodile breeding). The urban farms covered by detailed field research are indicated in Figure 4.2.

The second group of farms comprises eight facilities located in the central, well-connected districts of the city. Three of them are located in the south of the island in the Central Region, while five of them are in the northern part of the city on both sides of the Lower Seletar reservoir (four within the limits of the North Region and one in the North-East Region). Due to high building density, establishing urban farms in central parts of Singapore is difficult and requires adequate adaptive measures. For example, one of the analysed farms, which belongs to the Comcrop company, occupies the roof of a shopping centre – *SCAPE Mall, less than 200 m away from one of Singapore's main thoroughfares – Orchard Road within the Downtown area. The rooftop location means that the farm owners have to apply appropriate production methods, in this case – soil-free hydroponics. In another case, the owners of Citizen Farm, a company with its registered office in Queenstown in the southern part of Singapore, have foreseen the need to change the farm's location due to more and more intensive growth of building density in Queenstown and decided to use maritime shipping containers for plant production. Thanks to it, the farm is almost always fully mobile and prepared for the likely move.

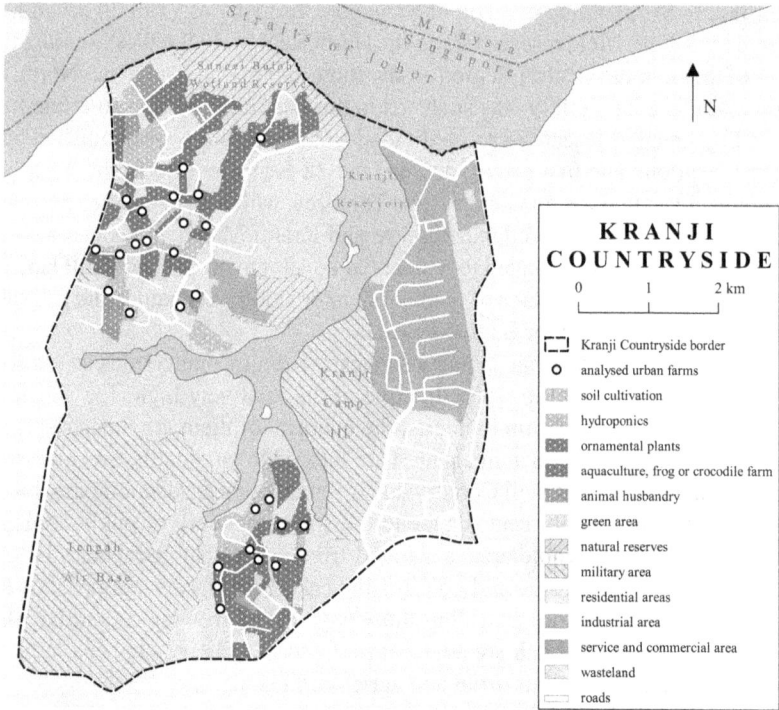

Figure 4.2 Forms of land use within the limits of Kranji Countryside.

Source: Górna, Górny 2021 (Creative Commons CC BY 4.0 Licence).

The need to change the location of many farms has proved one of the most serious challenges faced by contemporary urban agriculture in Singapore. During interviews with farm employees and owners, seven respondents emphasised that the basic problem they were facing was the approaching end of the lease period for the occupied plots of land, and thus the perspective of moving locations or liquidation. This mainly concerns farms that do not produce food on an industrial scale located within the limits of Kranji Countryside, for example, *Bollywood Veggies* and *Onesimus Garden*. The latter discontinued its operations after the end of the field research. However, facilities which are situated in advantageous locations are also facing a challenge associated with moving or liquidation, such as the three analysed farms located in the North Region, less than 500 metres from the MRT Khatib station within the Yishun functional sub-centre – *Oh Chin Huat Hydroponic Farm, Green Valley Farm* and *Pacific Agro*. All three farms were liquidated in 2019–2021 and the land they used to occupy has been allocated to residential buildings.

Until June 2016, the maximum permitted period of land lease for agricultural purposes in Singapore was ten years, with a possibility to extend the lease for ten more years. This was a significant impediment to operation of the farms, which were unable to achieve financial stability within such a short period of time. After a lot of contradictory information from representatives of the authorities and uncertainty reported by urban farmers associated with securing their livelihood, the AVA announced its decision to extend that period to 20 years (Tay 2016). However, it should be emphasised that it is the Singaporean government that makes the decision to extend the lease agreement and has full authority within this scope. Moreover, extension of the lease period by ten years does not mean that farms whose agreements expire in the upcoming time will manage to obtain an approval to extend them. According to the documents published on the website of the Singapore Land Authority, the institution competent for land management, the government's policy in this regard is clear. Although it is possible to extend the lease period, such a practice is rarely, outright sporadically applied, and each case is considered individually. For the authorities, the main arguments when it comes to regular review of ownership of municipal land are the insufficient land resources and the need to adapt the land use to the dynamically changing needs of the Singaporean economy. In the case of agriculture, the lease period may only be extended when the tenants have made substantial investments in the land or property and their activity is important from the point of view of "strategic national needs" (Singapore Land Authority 2023). Therefore, it means that farms which are insufficiently lucrative or do not fit in with the directions of development promoted by the authorities can be moved or – if they fail to obtain the rights to land use in another location – shut down. The SFA also offers a different form of lease – short-term lease of land for agriculture from one to three years (Singapore Food Agency 2020c). One of the farms taking advantage of such an agreement is the aforementioned mobile Citizen Farm located in Queenstown in the south of Singapore. Apart from the limited lease time, the authorities of the city-state can also control the distribution of agriculture through a system of licences for running a farm. Residents interested in such an activity, who have the right to use specific land, are obliged to submit a business plan to the SFA for a comprehensive evaluation. The foregoing practices show that the authoritarian government not only limits freedom and decision-making powers of Singaporeans in shaping urban space, but also their food sovereignty. The government inhibits the residents' initiatives within the scope of grassroots formation of the urban food system (in which they are actors) if those initiatives do not fit in with the chosen policy.

According to Singapore's spatial development Master Plan published in 2019, the only areas allocated to agriculture are those in the Lim Chu Kang district in the north-west part of the island (almost completely within the limits of Kranji Countryside), as well as the peninsula between the Tengah and Peng Siang Rivers and the south-east part of the Ketam Island, where aquaculture is currently pursued (Government of Singapore 2019). However, the total area of those locations is small, and more importantly – smaller than that currently occupied by urban farms. Therefore, after the end of the aforementioned lease period granted to individual farmers before 2019, some of the farms should be expected to wind down. Moreover, the Singaporean authorities had been planning to allocate considerable patches of land within Kranji Countryside to military purposes by 2021. The facilities

intended for relocation, including 14 urban farms analysed in the book (which occupy a total of 84.8 ha), were offered land with an area of 60 ha in the vicinity of the
Sungei Tengah reserve (Whitehead 2019) (see Figure 4.2). Taking into account the
fact that not all farms within Kranji Countryside were analysed in this study, one
should assume that the area of the land proposed by the government is insufficient
(since the area of the analysed farms alone is nearly 25 ha larger). Militarisation of
land previously occupied by agriculture is another form of legal and administrative
violence against urban farmers, who are unable to file an objection. Among the 36
analysed farms, only 13 (with a total area of 79.6 ha) are located in zones which
can be allocated to farming according to the Master Plan. The locations of the
remaining 23 sites, with a total area of 109.2 ha (including 14 within the limits of
Kranji Countryside), do not align with the provisions of the document, so they will
be either moved or shut down.

Although the situation of urban farms in Singapore is difficult and unstable, the
government's Master Plan 2019 includes an alternative for traditional cultivation,
which may become an opportunity for the development of agriculture in the city.
The document assumes utilisation of previously unused spaces for food production, including rooftops and building interiors (Government of Singapore 2019).
Such solutions are already being practised, among others, by two analysed farms
located in the central part of the island – *Citizen Farm* and *Comcrop*.

Figure 4.3 presents the period of operation of 37 analysed urban farms in Singapore, from the date of founding of the oldest one in 1974 until 2022. The number of
sites operating in the city during this time has been gradually increasing, exceeding
30 in 2010. The largest number of new farms was established in the 1990s. They
were located mainly within the limits of Kranji Countryside and within Khatib, in
the Yishun functional sub-centre. On the other hand, in the second decade of the
21st century, farms also began to appear in other districts of the city – in the Central
Region (three farms) and the North-East Region (one farm). Eight of the 37 farms
visited during the field research have been shut down so far. One of them (*Goodland
Hydroponic Farm*), located in Khatib, closed down in 2013. On the other hand, *Oh
Chin Huat Hydroponic Farms*, which were directly adjacent to it, were liquidated
in 2020, a few months after completion of field research. The three remaining farms
located in the area of Khatib – *Green Valley Farm, Pacific Agro Farm* and *Hua Hng
Trading/World Farm* – were also closed in 2021–2022. Three farms located within
Kranji Countryside have also discontinued their activity – *Onesimus Garden* shut
down in 2021, while *Sin Farms Trading Enterprise* and *Blooms & Greens Pte Ltd*
in 2022. On the other hand, the *Comcrop* farm, located on the roof of the *SCAPE
building several hundred metres from Orchard Road, was completely moved to the
Woodlands functional sub-centre in 2020. The examples of liquidated farms show
the significant impact of the changes planned by the authorities taking place within
the city's spatial and functional structure, on the distribution and activity of the
farms within the city.

The food policy of the Singaporean authorities has a precise focus – investing
in modern, intensive and profitable agriculture combined with the efficient use of
limited space. However, the activities of the national institutions ignore grassroots

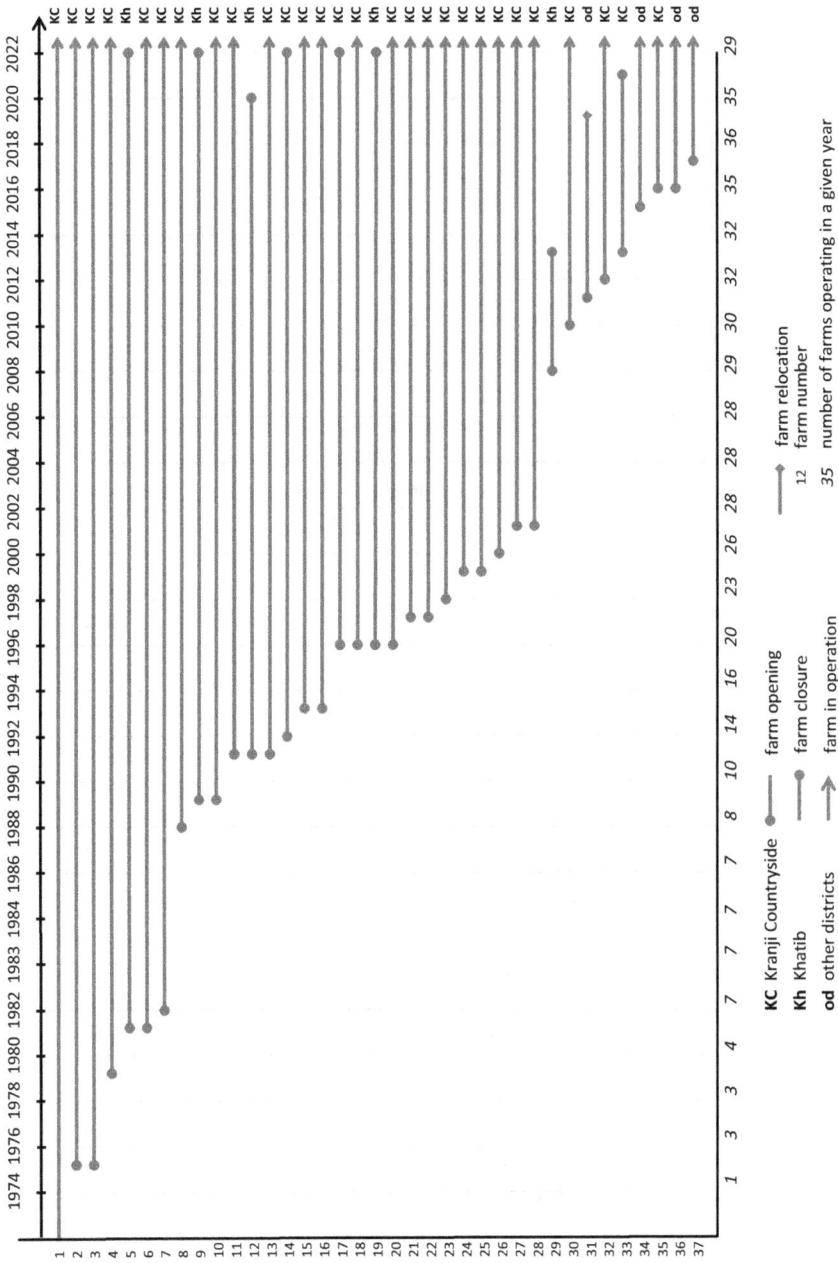

Figure 4.3 Period of operation of 37 analysed urban farms in Singapore in 1974–2022.

Source: Own study.

initiatives of the residents. According to the information collected during the interviews with representatives of two farms operating within the limits of Kranji Countryside (*Bollywood Veggies* and *Onesimus Garden*), these initiatives are based on the assumption that urban agriculture is more than monofunctional food production; it is also a multifaceted activity, where growing plants and breeding animals are just as important as building bonds between people, raising awareness and knowledge of organic cultivation and healthy nutrition, ecological education or care for the urban ecosystem. The fact that the authorities' approach is limited in the aforementioned regards has its consequences for the distribution of agriculture within the city space. It results in a small share of community gardens, which are typically present in the vicinity of housing estates or schools and kindergartens. Such gardens are typically established as a result of grassroots initiatives of groups of residents, which are limited and rarely tolerated by Singapore's authorities. Moreover, the government's policy towards agriculture also results in the majority of farms being concentrated in less economically valuable areas – on the edges of the island, in marshy and poorly connected locations. However, it turns out that as time passes and the city expands, even those areas become attractive to Singaporean authorities, and urban agriculture will be gradually moved to less and less advantageous locations, until it probably disappears completely in the end. Institutional and legal structures as well as centralised decision-making with regard to land use give the state full control over the spatial distribution of agricultural activity. Due to the short lease period and the fact that the authorities adopt their decision to extend it to changing economic conditions, distribution of farms in Singapore has been highly dynamic, in particular since the early 2020s.

4.5   Characteristics and functions of urban agriculture

4.5.1   *Structural and production characteristics*

The area of individual farms visited during the field research was estimated using the tools of Google Earth Pro. Account was taken of both the total area of each parcel and the area allocated strictly to production. In the majority of cases, similarly to Havana, these values differ significantly due to the large area occupied by auxiliary facilities, that is, warehouses, offices, farm buildings, access roads or – as in the case of a rooftop farm – elements of electrical and ventilation systems. The total area of the analysed farms amounts to 188.6 ha, with an average area of a single farm being 5.24 ha. The production area is almost half the size, amounting to 102.2 ha and with an average of 2.84 ha. These values are considerably higher than in the case of Havana, as discussed in the previous chapter, where the total area of the analysed gardens was 32.3 ha (with production area being half that size, similar to Singapore), while the average area of an urban garden amounted to merely 0.75 ha.

The research sample shows considerable differences in size between the group of 28 farms located within the limits of Kranji Countryside and the eight farms located in other districts of Singapore. In the case of the former, the average area of a single plot of land amounts to 6.38 ha, while the average production area amounts to 3.49 ha.

The group of farms in Kranji Countryside also includes the largest analysed farm, with an area of 21 ha (and production area of 14.85 ha). Due to the availability of spatial resources, agriculture in this part of Singapore takes up much larger areas than in other districts. The average area of farms located outside Kranji Countryside is merely 1.25 ha, with a production area of 0.56 ha. The share of production area in the total area of the parcel is lower in the central districts of Singapore (where it amounts to 45%) than in Kranji Countryside, where this share stands at 56%. The difference is due to the methods used in agricultural production. Central districts of the city utilise vertical crop cultivation; as a result, it is not necessary to use as vast areas of land as in the case of crops grown directly in the soil, which is the method applied in the outskirts of the metropolis. This gives rise to methodological issues since estimation of the production area was considerably more difficult for the four farms using the aforementioned vertical farming method: *Citizen Farm* and *Comcrop* in the Central Region, *Pacific Agro Farm* in the Khatib district and *Sky Greens* in Kranji Countryside. Although they take up less space in horizontal terms, their production area is larger thanks to the use of vertical farming. Moreover, in the case of farms that apart from outdoor soil and soil-free crop cultivation also use indoor hydroponics, such as *Kok Fah Hydroponic Farm, Eden Garden Farm* and *Orchidville*, it was impossible to estimate the aggregate production area exclusively on the basis of images from Google Earth Pro. Hence, the foregoing values for total and average area used by urban agriculture in Singapore should be treated as illustrative and subject to a certain margin of error.

The agricultural sector of Singapore is dominated by egg, fish and vegetable production to meet the local needs, as well as production of orchids and ornamental fish for export (Tey et al. 2009). According to official data published by the SFA, in 2019, the total number of farms holding the governmental licence stood at 220, of which as many as 109 were involved in marine aquaculture. Among the sites located ashore, 77 were involved in production of leafy vegetables (or leafy greens), that is, kale, spinach, cabbage, lettuce, gai lan, water spinach, amaranth greens or bok choy (also called Shanghai bok choy). Twelve farms dealt with inland aquaculture, six with poultry breeding and 11 with breeding other animals, such as goats (mainly for milk), frogs or shrimp. At six farms, bean sprouts were grown (Singapore Food Agency 2020b).

Similar to Havana, the overwhelming majority of the analysed farms – 28 (i.e. 77.1% of them) – base their activity exclusively on plant production. Six facilities, on the other hand, specialise in animal production only. Only in two cases (*Kok Fah Hydroponic Farm* and *Citizen Farm*) are plant and animal production combined, with animal production serving an ancillary role. Among the 30 farms whose activity involves plant production, 19 grow exclusively edible plants, while nine of them grow only decorative plants (mainly orchids). Two farms combine production of edible and decorative plants. The structure of food crops is similar at all analysed sites. The vast majority of the plants grown are the aforementioned leafy vegetables (21 farms, i.e. 100% of those engaged in food production, 70% of those engaged in plant production and 58.3% of all farms). Leafy vegetables, often called Asian leafy greens or Asian greens, are an important component of diet not only in Singapore itself, but throughout the region of Southeast Asia. They are also

Figure 4.4 Aquaculture of edible fish and koi fish at *Nippon Koi Farm.*
Source: Photograph by K. Górny, January 2019.

adapted to being grown in hot and humid equatorial climate, in places exposed to a lot of sunlight. The analysed farms typically grew the following plants: bok choy, bai cai – ten farms; lettuce varieties: Chinese lettuce, butterhead lettuce, Korean lettuce – eight farms; spinach varieties: amaranth spinach, baby spinach and red Ceylon spinach – six farms; and nai bai, kai lan and cai xin – also six farms each. Other groups of edible plants grown in Singapore include cherry tomatoes (two farms), lady's fingers or okra (three farms), cucumbers (one) and onion (one). Three farms (*Green Circle Eco Farm, Bollywood Veggies* and *Cactus Sunrise Community Farm*) also grow fruits in addition to vegetables, including bananas, guava, jackfruit, limes, noni, papaya or passion fruit.

Out of the six farms specialising in animal production, five are dominated by aquaculture (see Figure 4.4). Moreover, analysis of satellite images and an online query prove that there are many more farms involved in aquaculture within the area of Kranji Countryside. Their distribution is visible in Figure 4.2. During field research, it proved impossible to visit them and conduct any in-depth analysis due to the fact that they were inaccessible to outsiders. The situation was similar in the case of the poultry and crocodile farms.

In the analysed sample of 36 farms, ornamental fish (koi) were bred in three facilities, while two other farms bred both ornamental and edible fish. At two farms – *Jurong Frog Farm* and *Cichlid Aquarium Supplies* – fish farming accompanies the core activity of those farms, which is farming of American bullfrogs (see Figure 4.5). The meat of American bullfrogs is in both cases intended for sale in supermarkets,

Figure 4.5 Edible American bullfrogs bred at the *Jurong Frog Farm.*
Source: Photograph by K. Górny, January 2019.

while *Jurong Frog Farm* also supplies a restaurant serving frog meat dishes on site. The last farm dealing with animal breeding is *Hay Dairies*. The company was founded in the 1920s by Hay Yak Tang. At the time, it specialised in poultry farming but switched to pig farming 50 years later. The farm quickly became the main pork producer on the local market. However, in the 1980s, the authorities of Singapore introduced a ban on pig breeding in the city area. This forced the owners of *Hay Dairies* to adapt their activity to the applicable law. With the support of the Primary Production Department (subsequently transformed into the Agriculture & Veterinary Authority, and later on in today's SFA) the farm started breeding goats for dairy production. The first animals were imported from the United States. They belonged to the following breeds: Alpine, Nubian, Toggenburg and Saanen. The *Hay Dairies* farm currently owns 600 goats and sells dairy products on the Singaporean market.[5]

The analysed farms also include those characterised by a highly diversified production structure. An example of such a farm is *Citizen Farm*. Apart from growing leafy vegetables, such as amaranth spinach, mizuna, chard and Red Vein Sorrel, it also grows microgreens, that is, seedlings of vegetables and herbs in an intermediate form between sprouts and ripe vegetables. The microgreens grown at *Citizen Farm* include micro basil, micro coriander and micro rucola (salad rocket). Apart from plants, the farm also grows edible mushrooms of the pink and golden oyster mushroom species. What is more, *Citizen Farm* also breeds hens, but only for presentation purposes for visitors. An important aspect of the farm's operation is testing modern technologies in the field of sustainable food production. One of the owners' innovative initiatives is an insect farm. Breeding black soldier flies (*Hermetia illucens*) has two basic purposes. First of all, the insects transform organic waste into fertiliser used for plant crops. Second, their larvae are intended for consumption by both people and farm animals (e.g. hens and frogs).

The structure of agricultural production in Singapore also varies in spatial terms. Farms specialising in animal production are concentrated primarily in Kranji Countryside, far away from the centre. Due to the need for a large area of land and unpleasant odours, animal husbandry and aquaculture are located at a distance from housing estates. Central parts of the city only feature farms involved in plant production. The aforementioned *Citizen Farm* is an exception in that regard.

To sum up, the majority of the analysed farms specialised in plant production. They can be divided into two major groups. The first one comprises farms that grow decorative plants (various species of orchids) for export. The second group consists of farms whose activity is based on food production. Contrary to farms in Havana, the structure of their crops can hardly be considered varied. In all analysed cases, the edible plants grown are leafy vegetables. Although this group on its own is represented by many different species, it does not change the fact that Singaporean farms supply the local market almost exclusively with one group of plants – leafy greens. On the one hand, this may be considered the result of the residents' nutritional needs, that is, of the local market's demand for various types of lettuce, cabbage and spinach, which are typical of Asian cuisine. On the other hand, the prevalence of leafy greens may be due to the fact that they spoil quickly, so it is advantageous to grow them as close to the market as possible. The small share of plant crops from other groups means that local food production only supplies Singaporean stores with one group of products, while all others have to be imported.

As far as animal production is concerned, the situation is similar. Although there are farms that engage in goat, poultry, American bullfrog and crocodile farming in Kranji Countryside, Singapore as a whole is dominated by aquaculture. In the majority of cases, it is limited exclusively to breeding ornamental koi fish intended for the local and external markets. In the analysed sample, aquaculture was clearly dominant and only in two out of five cases does it include edible fish. Thus, the structure of animal production in Singapore demonstrates that urban agriculture meets an extremely small percentage of the needs of the local food market. Once again, the consumers' needs can be met almost exclusively with imported food.

4.5.2 Organisational and technical characteristics

As in the case of distribution of urban agriculture, its organisational and technical features are also controlled by the Singaporean authorities. Their policy with respect to the local food production was set out, among others, in the provisions of the Master Plan published in 2019 (Government of Singapore 2019). State support in this regard is addressed mainly to farms applying modern or highly profitable production methods. The policy of the authorities is reflected in organisational and technical features of the facilities visited during the field research. The choice of production methods or techniques is, on the one hand, frequently dictated by the financial incentives offered by the government, and on the other hand – limited by the legal provisions that prohibit specific solutions.

Figure 4.6 Semi-protected roofed cultivation directly in the soil at *Farm 85* in Kranji Countryside.

Source: Photograph by K. Górny, February 2019.

In spite of the authorities' efforts promoting modern production methods, at 14 out of 21 farms specialising in growing edible plants, the predominant method is still traditional soil cultivation (see Figure 4.6). Modern hydroponics is used in nine facilities. Other methods applied in Singapore included organoponics (see Figure 4.7), that is, cultivation in raised plant beds surrounded by borders (typical of Cuban cities, including Havana, which was described in the previous chapter) – seven instances, aquaponics – three instances, as well as cultivation in plastic or jute grow bags (two instances and additionally ten farms involved in growing ornamental plants). Moreover, in the analysed sample, in 11 instances (52.4% of farms specialising in food crops) more than one production method was used.

This is a good point to discuss the functioning of the two aforementioned systems of soil-free crop cultivation used in Singapore – hydroponics and aquaponics – as well as explain two other accompanying terms – controlled environment agriculture (CAE) and vertical farming.

Hydroponics, also known as water culture, is a technique of growing plants on water mediums (mineral nutrient solutions). Plant roots absorb nutrients directly from the water medium in which they are submerged. Excess water is filtered to a tank, from which it is later filtered back to the level of plant roots. Hydroponics enables cultivation of crops in areas without soil or where soil is unsuitable for cultivation due to pollution or insufficient fertility. A production method similar to hydroponics is aeroponics, which involves water and nutrients being sprayed onto the roots in aerosol form.

Figure 4.7 Crops grown in raised plant beds in one of the greenhouses at *Green Valley Farm* in Bah Soon Pah Road in the Khatib area.

Source: Photograph by K. Górny, February 2019.

Merle H. Jensen (1997) distinguishes two types of hydroponics based on the presence of the substance stabilising the plant roots – liquid hydroponics, where the roots are submerged exclusively in water rich in nutrients, and aggregate hydroponics, where the roots are placed in various forms of aggregates – sand, gravel or synthetic granulate. Hydroponics can operate in an open system, where the water medium is not reused after being fed to the plant roots, and in a closed system, where the medium surplus is collected, refilled and reintroduced into the system (Jensen 1997). The history of hydroponic cultivation dates back to the 17th century, when the first experimental soil-free cultivations appeared in France and England. In the following centuries, the method was gradually improved, and thanks to reduction of the costs of the system construction, it became more and more popular in commercial production (ibidem). Hydroponic systems are currently typically located in closed rooms or greenhouses (Thompson et al. 1998). They also take the form of CAE (see Figure 4.8). It is a technologically advanced and frequently capital-intensive approach to food production. Strict monitoring of the temperature, humidity and provided nutrients ensures optimal growth conditions at every stage of the plant's development. Sunlight is replaced by artificial heating and LED lighting (Jenkins 1997; Benke & Tomkins 2017). CAE is not a synonym of hydroponics; however, both methods are frequently used simultaneously (Jenkins 1997). In addition, hydroponic modules are often set up vertically, creating structures of up to several storeys high. This practice, referred to as vertical farming, allows for the crop area to be increased while simultaneously reducing the amount of space

Figure 4.8 Hydroponic cultivations in a controlled environment inside a maritime container at *Citizen Farm*.

Source: Photograph by K. Górny, February 2019.

taken (Benke & Tomkins 2017). It has become popular in many cities around the world, including in Singapore, which suffers from a space deficit.

Another soil-free production method, which was applied in three of the analysed Singaporean farms, was aquaponics. The term was coined using two words – aquaculture and hydroponics. Aquaponics is therefore a system integrating the two aforementioned methods – fish farming and soil-free plant cultivation (Diver & Rinehart 2000). Fish excrements placed in a water tank are decomposed by bacteria into nitrates and nitrites (nutrients), which are then pumped into the tank in which the plant roots are submerged. Excess purified water is returned to the fish tank.

Both hydroponics and aquaponics entail a number of benefits. If they operate in a closed or semi-closed system, they allow for water and energy consumption to be cut down on. Above all, they reduce the amount of space taken up, which is particularly important for agriculture located within cities. However, the examples of Singaporean farms show that keeping both systems operating optimally and achieving high crop efficiency involves a high risk of failure, is labour-intensive and requires continuous investments in improving the technologies used.

At two farms – *Orchidville* and *Kok Fah Technology Farm*, located in Kranji Countryside, both soil cultivation and hydroponics were used. However, in both cases, the respondents emphasised that modern hydroponics is merely a supplementary method used for demonstration purposes. It is not efficient enough yet to be the core of a farm's business. The owners only decided to implement hydroponics due to financial support of the authorities provided to facilities that utilise modern technologies; however, it is soil production that serves as the main source of income for both farms. Interestingly, at *Orchidville*, the situation is different in the case of aquaponics (which is the third production method used in this location). The integrated system proved much more efficient than hydroponics, so the owner decided to increase the area dedicated to it. Other respondents that also pointed out issues with maintaining and optimising the operation of hydroponics and aquaponics included representatives of *Citizen Farm* and *Comcrop*. They emphasised that although use of the aforementioned methods is promoted by the authorities, it is time-, labour- and capital-intensive, while optimisation of the production process requires a series of research, trials and experiments. What is more, both respondents emphasised that aquaponics in particular had proven to be a particularly unstable system and is sensitive to all types of failures, for example, power shortages or sudden changes in pH of water, leading to the fish dying off abruptly.

Two of the analysed farms stand out due to the multitude of production methods used. The first one is *Citizen Farm*, which uses soil cultivation, hydroponics, organoponics, aquaponics, mushroom farming, as well as poultry and insect farming. However, this wide range of forms of agricultural activity is largely demonstrative, educational and experimental. The farm's activity includes workshops dedicated to sustainable food production organised for the residents of Singapore. Apart from that, the farm engages in research works aimed at improving the applied modern farming methods. *Citizen Farm* is managed by a company called Edible Garden City, which – apart from running its own, multifunctional farm – also establishes urban gardens in various locations in Singapore. An example of such a garden is *Open Community Farm*, discussed in this study and operating next to a restaurant of the same name. One of the company's latest initiatives is installation of a garden on the roof of the Raffles City shopping centre in North Bridge Road in Downtown. During the field research, works aimed at transforming a recreational garden, which was originally located on the roof, into a vegetable garden were in progress (these works have already been completed) (see Figure 4.9).

The second case distinguished by a multitude of production applied methods is *Green Valley Farm* in Bah Soon Pah Road in the Khatib area. The farm was located on a parcel owned by the SFA and comprises 80 greenhouses leased by individual residents (see Figure 4.7). Each separated space used different production methods, depending on the personal preferences, including soil cultivation, hydroponics, organoponics and cultivation in pots and plastic grow bags. The farm, which in terms of its organisational form was reminiscent of allotments typical of European countries, was shut down together with other farms located in the Khatib area.

Information on composting was obtained from 11 representatives of farms specialising in plant production. In nine cases, compost was produced on site, using waste from the farm itself and neighbouring industrial plants (e.g. a brewery). *Citizen Farm* even used organic waste provided by the residents. In five instances, the open-air composting method was used, while at two farms, composting was additionally supported by vermiculture. Moreover, use of artificial fertilisers was not confirmed in any of the analysed facilities. However, their use is not prohibited in Singapore. According to the representative of the SFA,[6] this issue is not legally regulated. That is not the case for plant protection products. Each new product launched on the Singaporean market must be registered. Only two of the

Figure 4.9 A rooftop farm on the Raffles City shopping centre building under construction in 2019.

Source: Photograph by K. Górny, February 2019.

interviewed respondents – representatives of *Farm 85* and *Kok Fah Technology Farm* in Kranji Countryside – confirmed that artificial pesticides and herbicides are used at their farms, but only when it is necessary, for example, when the majority of crops suffer from a disease. In other cases, nets protecting the crops against insects laying eggs in the soil are used, as well as bags protecting ripening fruit against insects and birds or manual removal of pests from the plants (see Figure 4.10). Another popular, eco-friendly agent used in Singapore is neem oil.

Information on crop irrigation was obtained from respondents from 13 out of 30 farms dealing with plant production (or combining plant and animal production). At seven of them, use of drip irrigation system was confirmed, at two – manual watering, at two other farms – both drip irrigation and manual watering and at one – sprinklers. Four of the analysed farms used water from the pipeline for irrigation (*Comcrop* farm located in Downtown and three farms located in the Khatib district). At three other farms, located in Kranji Countryside, rainwater collected in special ponds was used to water the plants. Representatives of farms using rainwater emphasised that such a solution limits water consumption and makes the farm independent from the costly resources of municipal waterworks. Singapore is located in the tropical rainforest climate zone (according to Köppen climate classification), characterised by high total annual precipitation, which is intense in every month of the year. However, due to the inefficiency of the waterworks system, the city-state is struggling with supply of water suitable for food production.

Figure 4.10 Semi-protected cultivations at *Kok Fah Technology Farm*. The photograph features yellow flypaper.

Source: Photograph by K. Górny, January 2019.

This is why collecting rainwater improves self-sufficiency of farms. In the remaining cases, water was taken from a well.

Fourteen farms utilised structures with impermeable and semi-permeable sheets
for protected and semi-protected cultivation. Their purpose was to protect plants,
especially more sensitive edible plants, from torrential rains and excessive solar
radiation (see, for instance, Figures 4.6, 4.7 and 4.10). Assembly of protective
structures allows for more control over the quantity of water reaching the plants
at subsequent stages of their growth; however, it also requires use of artificial
irrigation.

Singapore, as a highly economically advanced country, also promotes itself as
a country that cares for the natural environment. Meanwhile, Singapore's government is still reluctant to introduce not only a complete ban on plastic bags, but
also payments for such bags, which are common in countries with similar and
lower income per capita (Channel New Asia 2019). In spite of having signed the
Resource Sustainability Bill (*The Straits Times* 2019) – an act regulating the issues
of waste recycling and reuse by large producers – in 2019, the Lion City still serves
as a negative example of a metropolis that struggles with the problem of excessive
use of disposable plastic packaging. This problem was also noticed during field
research. At the overwhelming majority of farms specialising in plant production,
food is wrapped in disposable plastic packaging. Taking into account the fact that
the majority of farms sell their products on the local market, use of plastic bags
and boxes is unnecessary and harmful to the natural environment as well as the

Figure 4.11 Food in plastic packaging in a NTUC FairPrice supermarket.
Source: Photograph by A. Górna, February 2019.

consumers themselves. Excessive use of petroleum packaging concerns not only locally produced, but also imported food. The pervasiveness of plastic on shelves in Singaporean stores stands in striking contrast to the image of a green city created by the authorities (see Figure 4.11). This contradiction is all the more visible when considering the data on waste recycling published in 2020 by Singapore's National Environment Agency. In 2019, 34% of recyclable waste produced in Singapore was exported to neighbouring countries. Although the percentage of recycled waste among the 66% that was not exported amounted to 59% in Singapore, it applied largely to waste such as rubble (99% of rubble mass was recycled), ferrous metals (99%), non-ferrous metals (99%) or used slag (98%). Meanwhile, recycling of plastics, mainly light plastic, which is harmful to the natural environment and whose volume is larger than that of rubble or metals, stands at 4%; recycling of glass at 14%; biological waste at 18%; and paper at 44% (National Environment Agency 2020).

The overwhelming majority of farms – 34 out of 36 – is run by private companies. The two remaining cases are a community garden – *Cactus Sunrise Community Farm* (see Figure 4.12), located in the vicinity of a residential apartment building in the North-East Region, and *Green Valley Farm*, located in the Khatib area, which comprises a number of greenhouses leased by individual residents (see Figure 4.7). Information on employment was obtained from 17 respondents. Among 16 facilities, the average number of employees per farm is 25 (this number was nearly six times lower in Havana). *Kok Fah Technology Farm* employs the largest number of people – 80. Other farms where the number of employees exceeds 25 are: *Qian Hu Corporation* (60), *Oh Chin Huat Hydroponic Farms* (50)

Figure 4.12 The community garden *Cactus Sunrise Community Farm* located directly in front of a residential building. Raised plant beds are used in the garden.

Source: Photograph by A. Górna, January 2019.

and *Citizen Farm* (40). The remaining facilities employ fewer than 25 people. *Open Community Farm*, which is a garden owned by a restaurant of the same name, only has one employee delegated from the *Edible Garden City* project. The majority of farms employ full-time workers. Only at *Onesimus Garden* are the people working on crops paid for the specific work performed, for example, weeding a plant bed. The remaining farms have permanent employees. In one instance, the farm hires seasonal workers during harvest. The facility which was omitted from the foregoing estimates is the aforementioned *Green Valley Farm*. Since all greenhouses were leased by individual residents, the farm did not employ any workers. A single greenhouse was usually taken care of by one to two people living in the immediate vicinity (typically pensioners from nearby housing estates). With 80 greenhouses, the number of people involved in the operation of *Green Valley Farm* can therefore be estimated at 80–160.

An interesting aspect related to the employment structure is the employees' nationality. Foreigners – mainly Chinese, Malaysians, Indonesians, Bangladeshi and Indians – are employed in nine out of 36 farms. In five cases, they made up 100% of the workforce, while in three cases – more than a half of it. At *Citizen Farm*, out of 40 employees, one came from Taiwan, while the remaining were Singaporeans. The high share of foreigners among the farm employees is due to the fact that as a workforce, they are much cheaper than citizens of Singapore, thanks to which employing them is much more profitable from the point of view of farm owners. However, they do not have full discretion since the state authorities have also introduced significant restrictions in this regard. First of all, farms registered with the SFA can only hire employees from three groups of source countries: a traditional source – Malaysia; North Asia sources – South Korea, Hong Kong, Macau, Taiwan; non-traditional sources – Bangladesh, India, Myanmar, the Philippines, China, Sri Lanka, Thailand. Second, the time period for which foreigners can be employed has also been limited, and amounts to 14 years for unqualified workers and 22 years for qualified workers (Singapore Food Agency 2020a). What is more, according to the respondent – a representative of the *Bollywood Veggies* farm (located within the limits of Kranji Countryside), in line with the government's recommendations, the number of foreign employees cannot exceed the number of domestic employees. This practice is meant to limit the possibility of importing cheap workforce from neighbouring countries and support the local labour market, thus reducing unemployment.

In four instances, the employees (both foreigners and Singaporeans) are provided with an option to live at the farms in special designated apartments. This practice only applies to facilities located in Kranji Countryside – *Jurong Frog Farm, Bollywood Veggies, Onesimus Garden* and *Hay Dairies*. It is due to the fact that the aforementioned farms are located away from residential areas, which limits the possibility of daily commute to work. The opportunity to live at the workplace also leads to considerable reduction of the costs of living, which can be a factor when it comes to living in Singapore, especially for the foreigners. At three farms (*Jurong Frog Farm, Bollywood Veggies* and *Hay Dairies*), it is the foreign employees – Malaysians, Bangladeshis and Indians – that live on site. On the other hand, *Onesimus Garden* employed Singaporeans (senior citizens and former prisoners) who had serious difficulties with finding a job and paying for their own accommodation. Thanks to being able to live at the farm free of charge, excluded persons were included in the local labour market, which enabled them to make a living. In the remaining 13 cases, employees live close to the farm or commute from other districts of Singapore using transport organised specifically for this purpose.

During the conducted interviews, 14 respondents referenced the support offered to urban farms by Singapore's national authorities and institutions. In seven instances (*Kok Fah Technology Farm, Orchidville, Nippon Koi Farm, Oh Chin Huat Hydroponic Farms, Pacific Agro Farm, Green Valley Farm* and *Comcrop*), they admitted to having used the support of the AVA (currently the SFA). The assistance mainly involved conducting expert analyses, testing modern technologies and testing the quality of the produced food. In turn, representatives of the remaining seven farms emphasised that they did not use the authorities' support because they did not need it or it was too costly (it works based on loans or credits). According to representatives of *Bollywood Veggies* and *Onesimus Garden*, the state only supports highly profitable farms that use modern technologies. Their opinion is in line with what the Executive Manager of the Food Supply Resilience Group, an organisational unit of the AVA, said in the interview. He stressed that Singaporean authorities are committed to promoting efficient, commercial agriculture of industrial nature, based on high-tech, as well as production in a strictly controlled environment that enables generation of the highest profits possible. For this reason, farms such as *Bollywood Veggies* and *Onesimus Garden*, which function largely based on educational and social activity and whose production methods should be considered traditional, do not align with the model of modern urban agriculture promoted by the authorities. Such units need to rely on other types of support (e.g. from people visiting the farms, guided tours for school trips, online fundraisers).

An example of a method of obtaining non-governmental funds is collaboration between producers as part of the Kranji Countryside Association. It brings together 40 farms operating not only in Kranji Countryside, but also in other parts of the city. The association is a platform for sharing experience, technology transfer, joint promotion of local production and agritourism, as well as coordinated measures aimed at improving public transport to and from the city centre (Henderson 2009). One of its initiatives was launching a shuttle bus line – Kranji Countryside Express, which departs from the nearest Kranji subway station. The bus runs between farms (mainly those offering educational and recreational activities), transporting their employees and residents. The cost of a single two-way trip is SGD 3; however, the bus line is not very popular. This is mainly due to the infrequent schedule and limited number of stops (Górny & Górna 2019). Only a small group of five farms (three of which will be subject to a detailed analysis in this chapter – *Bollywood Veggies, Hay Dairies* and *Jurong Frog Farm*) decided to take part in the Kranji Countryside Express undertaking. This prevents them from collecting funds sufficient for marketing activities that would allow for more effective promotion of the line. The Kranji Countryside Association also regularly organises organic food markets – Kranji Countryside Farmers' Market, which local producers and consumers from all over Singapore participate in on an annual basis.

The conducted interviews demonstrated that owners and employees of farms in different locations know each other and regularly cooperate as part of the Association. An example of this is use of waste from a poultry farming facility, *Seng Choon Farm* (included in Figure 4.2), for compost production in *Onesimus Garden* and *Farm 85*. Cooperation between the farms, which the Association aims to gradually expand, can provide the basis for a sustainable, self-sufficient system that allows for limiting resource losses.

Apart involvement in the activity of the aforementioned NGO, other forms of the residents' activity within the scope of urban agriculture are extremely limited. The government of Singapore has a restrictive approach to civic activism (Tan & Neo 2009). It is not conducive to the development of community gardens, which are an expression of the residents' grassroots initiative. Any activity of this type in Singapore is subject to an extremely restrictive institutional framework, where all people involved have predetermined roles and functions. Lack of freedom to act makes it difficult to develop interpersonal bonds and leads to disintegration within an organised group of residents (ibidem). In 2005, the authorities of Singapore launched the Community in Bloom programme, which supported the establishment of community gardens within the city. Only one of the Singaporean facilities analysed in this book is a community garden – *Cactus Sunrise Community Farm*. Another example of activity which includes individual initiative of the residents is *Green Valley Farm*, where interested individuals lease small greenhouses on a plot of land owned by the SFA. However, it turns out that this form of residents' initiative is not desirable for the Singaporean authorities as the farm has been shut down. Based on their research, L. H. H. Tan and H. Neo (2009) claim that the authoritarian rule in Singapore, which in this context relies on strong intervention in any organised grassroots activity of the residents, has a negative effect on the operation of community gardens, limiting their development opportunities as well as limiting the popularisation of the idea of civic activity itself. What is more, the authorities' actions deprive the residents of opportunities to shape the space around them, which, in turn, clashes with the vision of democratic municipal authorities that are open to the society, as presented, among others, by the geographer D. Harvey in his essay entitled *Right to the City* (2008). D. Harvey rightfully argues that the local population should be considered an important actor and decision-maker in the process of city development. However, Singapore in 2023 remains an undemocratic state; there are no hints that this is going to change in the nearest future.

4.5.3   *Product distribution and functions fulfilled*

Since 34 analysed farms are private companies, the food they produce is intended for sale. In two other facilities – *Cactus Sunrise Community Farm* and *Green Valley Farm* – food goes to the people involved in the farms' operations. Determining the methods of product distribution was possible for 23 of the analysed farms. For 17 of them, the product paths were determined on the basis of information provided by the respondents, while for six others they were based on field observations and an online query. Only two facilities involved in breeding ornamental koi fish – *Qian Hu Corporation* and *Nippon Koi Farm* – export some of their products outside Singapore, among others, to Malaysia, Cambodia, Indonesia and Vietnam. *Cichlid Aquarium Supplies* is also applying for the government's permit to sell fish abroad, but at the time of the field research, such a permit had not yet been granted.

An important characteristic of the analysed farms is diversification of the forms of food sales. In nine out of 23 cases, the products were distributed in more than one way. Sale on site was combined with supplying restaurants or supermarkets, while some facilities also had their own online stores. This way, the producers expanded their options of reaching a larger group of consumers, both those who wished to purchase supplies on site, frequently combining a visit to the farm with a tour of the facility, as well as those who shop in large stores or online. Brick-and-mortar stores and supermarkets all over Singapore were supplied with food from 12 farms, and in four of those instances, the entire production was allocated for sale to supermarkets located in various parts of Singapore. During the interviews, four respondents listed the largest Singaporean supermarket chain – NTUC FairPrice – as the main buyer of their products. The company owns a chain of 100 supermarkets and large grocery stores throughout Singapore. The second chain, mentioned by the representative of *Hay Dairies*, a farm specialising in production of dairy products from goat milk, was PRIME Supermarket. This family business owns 24 supermarkets in the city. Due to the large number of stores supplied with food from the aforementioned 12 farms, as well as variable availability of products in individual locations, it was impossible to describe the exact route each type of goods took. The exceptions are vegetables and herbs from the *Comcrop* farm. Basil and mint produced here are sold in the RedMart online store, while lettuce is sold in NTUC FairPrice supermarkets in four locations: Woodlands, Woodgrove, Upper Thomson and Bishan Junction 8. The distance between them and the current registered office of the farm in the Woodlands district is, respectively, 4.7 km (2.95 km in a straight line) (the Woodlands store), 5.1 km (3.91 km in a straight line) (the Woodgrove store), 18 km (10.9 km in a straight line) (the Upper Thomson store) and 20.8 km (11.9 km in a straight line) (the Bishan Junction 8 store). These four supermarkets are located in the northern part of the Central Region and, just like the farm itself, in the North Region. Thus, *Comcrop* limits the sale of its products to the northern and north-eastern part of the island.

Apart from the supermarkets, eight out of 23 farms also supply Singaporean restaurants. In five instances, the gastronomic facilities operate within the farms or in their immediate vicinity. Those five sites, apart from providing food, also conduct gastronomic activity. In the three remaining cases, the restaurants are situated in various locations in Singapore.

Nine farms sell the food they produce on site. The sale is conducted both during the week and at weekend markets organised by the farms. Distribution of crops at the place of their production typically shortens the supply chain and allows for direct interactions between the producer and the consumer. In the case of Singapore, however, the implications are more serious than in Havana and Kigali. Due to large distances and peripheral location of the majority of the farms, they are not the everyday shopping destinations for Singaporeans since they need to overcome long distances to reach most facilities, especially those located in Kranji Countryside. Moreover, the sparse public transport network in this part of Singapore necessitates travel with a private means of transport – a car. Therefore, the positive effects of shortening the product path and direct contact between the producer and the consumer are diluted by the negative impact of increased car traffic on the natural environment.

Eight of the 23 farms have their own online stores. Residents can order selected products, which are subsequently transported to their homes or pickup points. Once again, this increases car traffic and thus air pollution caused by exhaust emissions. Not only the use of plastic packaging, but also the peripheral location of the farms and the need to transport food for long distances counters the benefits gained from typical urban agriculture in other metropolises of the world, including in two other cases described in this book.

Nevertheless, some of the analysed facilities fulfil functions which can be considered positive from the perspective of the natural environment and the society. These include, for example, reuse of municipal waste through local compost production, as well as measures aimed at ecological education and support of excluded social groups (seniors, former prison inmates, psychiatric patients, lonely and disabled persons) by involving them in work at the farms. Four respondents, representatives of *Bollywood Veggies, Citizen Farm, Onesimus Garden* and *Comcrop*, confirmed that apart from food production, social activity is also an important aspect of the farms' operation. The aforementioned facilities provide employment to the elderly, the disabled or former prisoners, who would not be able to find employment on the Singaporean labour market. Moreover, 12 of the analysed farms are accessible to people who wish to explore them: eight among them offer organised tours (educational or integrative), while two offer trainings and workshops. In this regard, there are two particularly noteworthy farms whose educational and/or recreational functions are dominant and much more important than the production function itself (*Citizen Farm* and *Bollywood Veggies*) as well as two farms where educational and recreational activities are meant to stimulate the sale of food products (*Hay Dairies* and *Jurong Frog Farm*) (Górna & Górny 2019).

The educational offer of *Citizen Farm* is thematically adapted to the requirements of various age groups. It includes organised tours and workshops within the scope of organic cultivation, composting and even hen farming. They are addressed to people interested in modern technologies used in contemporary agriculture, who wish to learn about the food production process in urban conditions, as well as visitors who are planning to start their own, private gardens. The farm can host up to 200 visitors at once. They are mainly Singaporeans; however, foreign tourists also sporadically show up. The services offered by *Citizen Farm* are so attractive that tours and workshops have to be booked far in advance due to the limited number of available spots (ibidem). The demand for similar activity among the citizens of Singapore is therefore considerable, which may be conducive to the establishment of farms with a similar business profile in the city.

In the case of *Bollywood Veggies*, the crop fields are also mainly demonstrative. The farm's activities involve first and foremost educational, social and gastronomic initiatives. The activities offered include a tour of the farm, planting workshops, rice cultivation workshops, culinary and art workshops, as well as outdoor and group games. A wide selection of attractions encourages visitors to spend the entire day at the farm (ibidem). According to the respondent, *Bollywood Veggies* is typically visited by organised groups and most of the farm's offer is addressed to a larger number of participants.

The offer of *Hay Dairies* includes free unsupervised exploration of the farm or paid organised educational tours with a guide. They are addressed to different age groups, and the key attractions include observation of goat milking and feeding in the morning. Moreover, visitors can purchase souvenirs associated with the farm's activity. Interacting with animals and being able to learn about the milk production process is an attractive option for city residents, who lack similar experiences in their everyday lives. Approximately 80% of visitors are Singaporeans. The farm's activity is based primarily on dairy production; however, the educational and recreational offer is an attempt at diversifying the sources of income. It is also a marketing tool, aimed at encouraging the visitors to buy the brand's products, including fresh goat milk (ibidem).

The last example is the American bullfrog farm – *Jurong Frog Farm* (Figure 4.5). In 2014, its owners decided to diversify its business and expand the offer to include educational activities. They currently include themed tours addressed to children, youth as well as adults. Most offers are intended for groups of 20–150 people.

The main attractions include educational talks, frog feeding, frog meat tasting, frog catching or taking photographs with a frog on one's head (*sic!*). The farm also offers other activities addressed to small groups of several people and individuals. At the farm, there is a souvenir shop and a small restaurant serving frog meat dishes. The farm's infrastructure allows it to host 150 people at a time, so similar to *Citizen Farm*, it has huge recreational potential (ibidem). According to the respondent – the farm's employee – the majority of visitors are Singaporeans, primarily school groups, church communities and groups of employees of various workplaces.

Taking into account the difficult financial position of many of the analysed Singaporean farms, income diversification can serve as an opportunity for them to remain on the market. Expanding their activity to include educational or recreational services can stimulate sales and, at the same time, increase income. Field research confirmed that six farms (*Oh Chin Huat Hydroponic Farms, Citizen Farm, Jurong Frog Farm, Hay Dairies, Onesimus Garden* and *Kok Fah Technology Farm*) organised paid tours (such information was not obtained in the remaining cases). The cost of the attractions ranges from SGD 3 per person (*Oh Chin Huat Hydroponic Farms*) to SGD 90 per person (*Citizen Farm*). The development of the recreational or tourist function among farms located in Kranji Countryside is supported by the Kranji Heritage Trail, which opened in 2011. It includes 14 historical sites and sites associated with agricultural production. The points on the trail include sites related to World War II (e.g. the Kranji War Memorial), ceramic industry (Thow Kwang Industry), as well as the Sungei Buloh Wetland Reserve (ibidem). The Kranji Heritage Trail includes five farms, three of which are analysed in detail in this work (*Bollywood Veggies, Hay Dairies* and *Jurong Frog Farm*). Information boards were placed next to each site on the trail to make it easier for residents to explore them on their own. Kranji Countryside, due to its location away from the centre of Singapore and thus away from frequented tourist trails, currently fulfils mainly agricultural, industrial as well as military roles. The trail was intended to turn the district into an active tourist destination. However, due to the lack of convenient commute options, few tourists have been visiting the farms located in that area so far.

Multifunctionality is a characteristic trait of urban agriculture and is considered one of its biggest advantages. Although in Singapore's case, supplying the population with locally produced food should be a priority in the authorities' policy in terms of increasing food self-sufficiency of the entire city, the social functions may prove equally important. Due to the small share of agricultural areas in the total area of the Asian city-state, its residents lack exposure to the production process of the food they consume on a daily basis. Therefore, maintaining urban farms and gardens that play an educational and informational role in Singapore's space can contribute to improved knowledge and awareness of organic agricultural methods and techniques, in particular among children. Moreover, involvement of excluded members of the population, such as senior citizens or disabled people, in the operation of individual facilities is an example of inclusive measures which are conducive to building an integrated society. Supporting development of functions other than food supply could present a chance for many farms analysed in this chapter to remain in business, which would simultaneously entail a number of benefits for the residents. Although some of the analysed facilities strive to diversify their income as much as possible, improving their financial standing, without governmental permits, in particular land use permits, they are unable to remain on the market. The current policy of the authorities is aimed at effective use of space and only efficient, highly profitable agriculture based on modern technologies receives governmental support. In their actions, Singaporean decision-makers fail to take into account the needs of the residents, which cannot always be met exclusively by securing financial gains.

4.6   Selected examples of urban farms

The following case studies describing four urban farms in Singapore differ both in terms of methods applied and in terms of organisational characteristics of the production process. The selected facilities are situated in various locations – *Oh Chin Huat Hydroponic Farm* within the Khatib functional centre, *Comcrop* in Downtown, *Onesimus Garden* on the Lim Chu Kang peninsula (Kranji Countryside) and *Kok Fah Technology Farm* on the peninsula between the Tengah and Peng Siang Rivers (Kranji Countryside). The farms can be divided into two groups, reflecting diametrically different situations of urban agriculture in Singapore. The first group includes *Oh Chin Huat Hydroponic Farm* and *Onesimus Garden*, which have been shut down due to being situated in a location that was too advantageous (*Oh Chin Huat*) or not fitting in with the policy directions of the Singaporean authorities with respect to local food production (*Onesimus Garden*). The second group includes two farms which are doing well, receive state support and are able to remain on the Singaporean market thanks to their use of modern production methods and high efficiency. Each case is accompanied by a diagram presenting the spatial layout of all facilities located within individual farms.

4.6.1   Oh Chin Huat Hydroponic Farms

Oh Chin Huat Hydroponic Farms is one of eight farms that have been shut down or moved since the time of the field research (January 2019). Due to the anticipated end of the 20-year lease period in 2021, the owners decided to shut down the farm because of the lack of perspectives for extending the lease. The interview was conducted with one of the farm employees. He provided information on the farm's operation as well as problems associated with the loss of support of the Singaporean authorities.

Oh Chin Huat Hydroponic Farms was located in Bah Soon Pah Road in the north of Singapore, in the south-eastern part of the Yishun functional sub-centre, more specifically at the Khatib housing estate, less than 640 m from the MRT Khatib subway station. From the east, *Oh Chin Huat* bordered *Pacific Agro Farm* (also covered by analysis in this work and also shut down due to construction of a housing estate) and the headquarters of the Agriculture and Veterinary Authority (currently the SFA), from the west – a gardening store called Pioneer Garden, and from the south – *Goodland Hydroponic Farm*, which had been shut down even before the field research. On the other side of Bah Soon Pah Road, there were two other farms included in this study – *Green Valley Farms* and *World Farm* – as well as a gardening store, Hua Hng Trading (all those facilities have been liquidated). In close proximity to the parcel occupied by *Oh Chin Huat*, there are three football stadiums, two golf courses, a military base – Khatib Camp – as well as a number of aquarium stores. The area of the Yishun functional sub-centre is characterised by regular, relatively densely packed service and residential buildings, predominantly apartment blocks of more than ten storeys. The *Oh Chin Huat* farm was located approximately 300 m from the closest residential buildings.

The farm was established in 1991, and throughout the period of its operation, it was financially and technologically supported by the neighbouring AVA, which allowed it to develop an efficient, highly productive hydroponic system. The farm occupied a plot of land of 2.4 ha and had 228 hydroponic greenhouses (Figure 4.13). They were arranged in six rows, two of which, in the northern part of the farm, were handled by the NFT (Nutrient Film Technique) system, developed in 1970 in

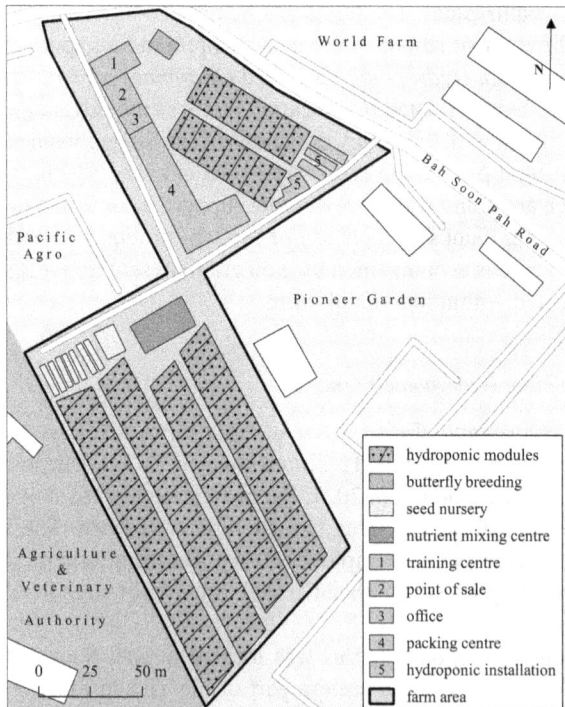

Figure 4.13 A diagram depicting the use of the parcel where *Oh Chin Huat Hydroponic Farms* were located.

Source: Own study on the basis field research.

Figure 4.14 Hydroponic cultivations in the NFT system.
Source: Photograph by K. Górny, January 2019.

Great Britain by Dr. Allen Cooper (Kao et al. 1991) (see Figure 4.14). The system involves a low level of water flowing through individual hydroponic modules. On the other hand, the four rows of greenhouses in the southern part of the farm used the DRF (Dynamic Root Floating) system (see Figure 4.15), which has been the subject of experimental works in Taiwan since 1987 (ibidem). The system enables year-round cultivation of leafy greens and herbs and is adapted to intertropical conditions. All *Oh Chin Huat* greenhouses were partly or fully covered with roofs. Semi-permeable and impermeable roofs protected the crops against excessive rainfall and solar radiation. Moreover, the structure of the greenhouse allowed for additional reduction of heat and moisture losses, which improved production efficiency.

Within the plot of land, there was also a drip irrigation system hub, a seed nursery, a warehouse for storing mediums used in the hydroponic systems, a butterfly pavilion and four outbuildings – a training centre, a sales point, an office and a packing centre.

Oh Chin Huat specialised mainly in the production of leafy greens (including Asian varieties of lettuce and spinach), herbs (basil, mint, coriander), cherry tomatoes, Japanese cucumbers and okra. Organic waste from the cultivation was not reused and no herbicides or fungicides were applied in the production process. Whenever pests or fungi appeared, it was necessary to remove an entire hydroponic module. Only nets were used for plant protection, which resulted in partial loss of crops.

Despite inclusion of a sales point, all products were packaged on site (in plastic) and transported to the NTUC FairPrice supermarket chain. Therefore, the short

Figure 4.15 Hydroponic cultivations in the DRF system.
Source: Photograph by K. Górny, January 2019.

distance between the farm and residential areas was insignificant for its location since the residents were unable to make purchases on site. However, local residents could participate in two-hour-long themed tours offered by *Oh Chin Huat*, dedicated to hydroponic production and growing herbs and spices. The cost of each tour was SGD 3 per person. What is more, as part of its activity, the farm organised trainings on growing your own food. *Oh Chin Huat* employed 50 people, the majority of whom – 30 – were office workers and facilitators of the aforementioned trainings. The remaining employees were in charge of the crops and product packaging.

During the interview, the respondent emphasised that since it was impossible to move *Oh Chin Huat* to a different location, the farm would be shut down, which took place before the end of the 20-year lease period (which only expired in 2021). Despite employing 50 people and having developed an efficient hydroponic system, perfected for nearly two decades, the farm was liquidated. The actions of the authorities are thus inconsistent since the AVA, a national institution, had provided financial and technological support to *Oh Chin Huat* since the beginning of its existence in that location. Shutting down the farm amounts to wasting the funds invested in production process optimisation and improvement of the hydroponic systems. However, the respondent emphasised that the authorities' decision was not based on unviability of the farm's activity, as the company' operations were flawless in that regard, but from the change in functional allocation of the entire district. The location of *Oh Chin Huat*, similar to all other facilities specialising in agricultural production in its vicinity, is soon to be allocated to housing development. The farm was officially shut down on 30 September 2020 and the parcel it occupied was completely cleared out. The hydroponic equipment, the tools as well as seeds and mineral media were disposed of as part of a yard sale.

4.6.2 Comcrop

The second farm analysed in detail, which similar to *Oh Chin Huat* no longer operates in its original location, is *Comcrop*. The company was established in 2011 and operated as a community garden during the initial period of its activity. Then, it was moved to the roof of one of Singaporean shopping centres – *SCAPE (see Figure 4.16), located in Orchard Link, a street running perpendicular to one of the main thoroughfares in Singapore – Orchard Road. The *SCAPE building is located in the Central Region, more specifically in the Central Area, informally known as The City. The area is dominated by high-rise commercial and service buildings with a lot of urban greenery. Orchard Road and its surroundings, due to the presence of luxurious stores and restaurants, is one of tourist attractions of Singapore and a place where residents can spend their free time. Apart from three commercial buildings – Ngee Ann City, Mandarin Orchard Singapore and Cathay Cineleisure Orchard – in immediate vicinity of the *SCAPE buildings, to its west, there is also a vast park of more than 2.4 ha.

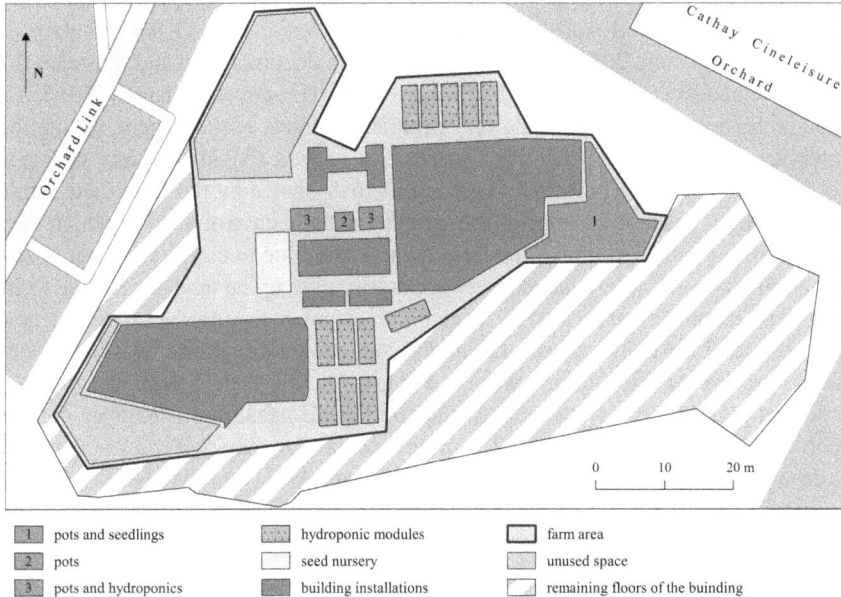

Figure 4.16 A spatial diagram of the *Comcrop* farm located on the rooftop of the *SCAPE building.

Source: Own study on the basis of field research.

The farm's location in that place required, first of all, a permit for leasing the roof, and second, a permit to turn it into usable space. The building owner is the Ministry of Culture, Community and Youth, the roof is leased by *SCAPE, while *Comcrop* subleased it until 2020. In 2018, the farm's owners decided to expand their activity and commence production in a second location – on the rooftop of Woodlands East Industrial Estate. The building is located in the north part of Singapore within the North Region. Due to maintenance works on the *SCAPE building and higher production efficiency in the new location, in 2020, *Comcrop* decided to liquidate its original seat and fully relocate to Woodlands. At the time of the field research in January 2019, the farm was operating in both locations. The interview was conducted with a *Comcrop* employee in charge of education and contacts with third parties. The following analysis will concern mainly the farm which no longer operates in its original location.

The area of the rooftop of the *SCAPE building is 0.28 ha; however, *Comcrop* only utilised a part of it, amounting to 0.14 ha. The remaining space was unused or allocated to the building's air conditioning systems. Within the farm, there were 12 A-shaped hydroponic modules (so-called A-frames) just under 4 m high (see Figure 4.17). The technique used here by *Comcrop* was most likely NFT; however, the respondent refused to provide information on the production technology in fear of it being used by the competition.

During the initial phase of its activity, apart from hydroponics, *Comcrop* also used aquaponics, which is an example of an integrated system combining aquaculture with hydroponics. Works on improving the production process based on the aquaponic method were presented in a Discovery Channel documentary. It was released in June 2016 on YouTube, where the documentary, and thus also the company, became very popular (2.7 million views) (Discovery Channel 2016). Although the documentary presents very promising effects of the system, during the interview, the respondent admitted that it had proved sensitive to any changes. As a result of a single power shortage, all the fish owned by the farm died, while the extremely unstable balance of the artificial aquatic environment in which they lived was disrupted. Although the farm owners continue to carry out research and experimental works, they decided to temporarily give up on aquaponics due to its time- and labour-intensity as well as incommensurably low production efficiency. At the time of the field research, the main production method used was hydroponics. At the *SCAPE location, it was used to grow herbs, such as basil and mint.

Apart from hydroponic modules, the rooftop also featured a number of pots with edible false shamrock (*Oxalis triangularis*) and rosemary. Products from the farm were used to supply NTUC FairPrice supermarkets (Woodlands, Woodgrove, Upper Thomson and Bishan Junction 8) as well as nearby restaurants. It was impossible to purchase them on site. Currently, the distribution network for *Comcrop* products has been expanded to include RedMart online stores.

One of the objectives of the company owners is to develop a sustainable food production system, which would be as self-sufficient as possible. Therefore, only manual forms of pest protection were used in crop cultivation, while the energy

Figure 4.17 Rows of hydroponic modules arranged on the rooftop of the *SCAPE building.
Source: Photograph by K. Górny, January 2019.

was obtained thanks to solar panels. That was also why the ultimately unsuccessful attempt at introducing an aquaponic system was made. Apart from food production, the farm also carried out a number of educational activities, such as trainings and workshops for children and youth. Moreover, apart from seven full-time employees, it also engaged the elderly and disabled persons. The height of some hydroponic modules was adapted so that less agile individuals could reach them comfortably. Involvement of excluded individuals was aimed at returning to the original, community-based nature of the farm's activity. *Comcrop* used the support of state institutions in terms of implementing modern production technologies as well as expert analyses concerning the quality of the food produced.

In its new location in the Woodlands district, *Comcrop* is gradually expanding its activity to include not only new species of plants grown and new product distribution locations, but also increasingly advanced production technologies. The company's website also emphasises the farm being a local business as it employs workers who live nearby. According to the authors of the online entry, this is meant to limit so-called food miles and commute miles (miles travelled by workers via means of transport) (Comcrop 2020). However, it should be noted that employing people who live nearby only reduces the distance they travel on their way to work. It has no impact on shortening the route of the products, which are transported to restaurants and supermarkets located even more than 20 km from the farm. Local sale could present a chance for limiting the so-called food miles; however, it is not practised.

4.6.3   *Onesimus Garden*

The first farm analysed in detail, which was located in Kranji Countryside, is *Onesimus Garden*. It was situated on the Lim Chu Kang peninsula in Neo Tiew Road. From the west, it was adjacent to *Seng Choon Farm*, which was involved in poultry breeding. To the east of it, on the other hand, there was an abandoned plot of land (see Figure 4.18). Two other farms included in this study – *Bollywood Veggies* and *Green Circle Eco Farm* – were also nearby. *Onesimus Garden* closed down in 2021, as indicated by analysis of satellite images – the parcel was cleared out and all forms of crops were removed.

Onesimus Garden was established in 2013 by members of a Christian organisation, whose name references a character from the New Testament – Saint Onesimus, a disciple of the apostle Paul of Tarsus. The interview was conducted with the farm's director and founder. Its activity was based on supporting excluded Singaporeans, especially the homeless, the disabled, as well as pensioners and former prisoners, for whom it is difficult to find a job. During the field research, *Onesimus Garden* engaged 25 people living in dedicated buildings on the parcel. Providing them with accommodation was necessary both due to their limited financial capabilities and due to the peripheral location of the farm. People working in *Onesimus Garden* performed all tasks associated with the production process, from weeding, planting, clearing plant beds, through compost preparation, harvest, to packaging products ready for sale. Although the employees did not receive regular wages, they were paid for individual tasks, for example, they received SGD 5 for weeding

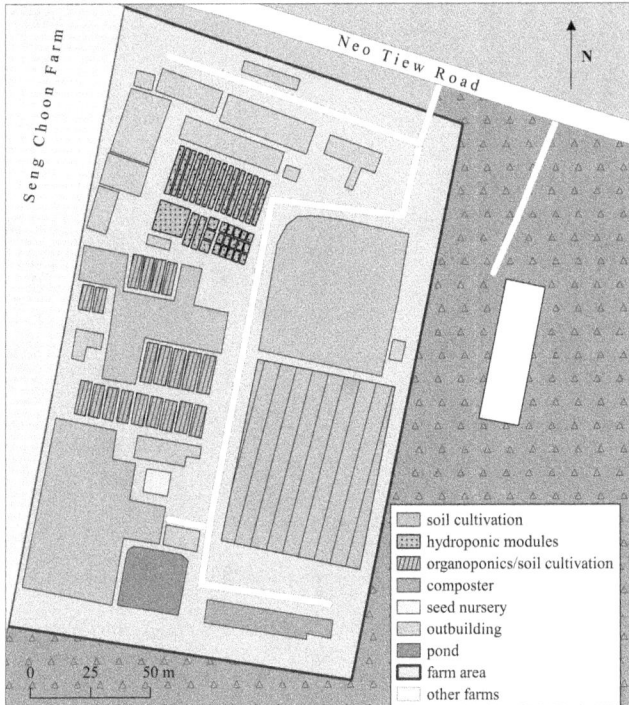

Figure 4.18 A diagram of *Onesimus Garden* located in Kranji Countryside.

Source: Own study on the basis of field research.

a single plant bed. According to the respondent, this form of work, developed during the first years of the farm's operation, proved the most efficient, especially in the case of former prisoners, for whom payment for specific work was the most motivating type of remuneration.

The farm occupied 5 ha, while its production area amounted to 1.43 ha. The parcel was divided into smaller sectors, where different production methods and technologies were used, predominantly semi-protected soil cultivation and organoponics (Figure 4.19). Moreover, in *Onesimus Garden*, there were 36 small roofed hydroponic modules owned by a third-party partner, *Grace Mission Hydroponic Farm* (see Figure 4.18). According to the information on the official website, in 2014, the farm also used aquaponics. However, during the field research, this method was neither observed nor mentioned by the respondent.

No artificial fertilisers were used at the farm; instead, compost was utilised. It was produced on site, using waste from a Singaporean cafe and brewery and with the support of vermiculture. Artificial pesticides were not used, either – they were substituted by manual labour and protective nets. If necessary, the crops were irrigated with water from a small pond on the edge of the plot; apart from it, rainfall was used in cultivation. The farm specialised in production of leafy greens,

Figure 4.19 Cultivations on raised plant beds covered by semi-permeable sheets. All structures were made of generally available materials.

Source: Photograph by K. Górny, February 2019.

predominantly pak choy, as well as other vegetables, such as okra. All products
were intended for sale. According to the respondent, the distribution network comprised Singaporean residents who were friends of the farm. They were the first re-
cipients of the products, which were subsequently distributed in an indirect manner.
Due to financial issues and insufficient workforce, the farm was already neglected
during the field research in 2019 and its infrastructure was damaged. *Onesimus
Garden* had remained on the market thanks to voluntary donations from visitors
taking part in organised tours, as well as sale of its products. However, due to low
profitability and social nature of its activity, the farm could not count on the authorities' support, which in the light of the new legal regulations put a serious question
mark on its future. Only the third-party partner, involved in hydroponics, received
funding from state institutions. However, those subsidies did not improve the situation of the farm as a whole. When comparing photographs published on the website
during the first years of operation of *Onesimus Garden* with the situation observed
during the field research, one can conclude that the farm had gradually deteriorated.
This was also confirmed by the respondent, who spoke about the country's policy
towards Kranji Countryside with a lot of resentment. In the face of the lease term
for the occupied parcel ending in 2021, the farm was shut down since the authorities
issued a negative decision regarding its extension. *Onesimus Garden* is an example
of failure to adapt to dynamically changing conditions of conducting business activity in Singapore. Moreover, its situation is also a symptom of the state's policy of
abandoning less lucrative urban agriculture that fulfils a clearly social role.

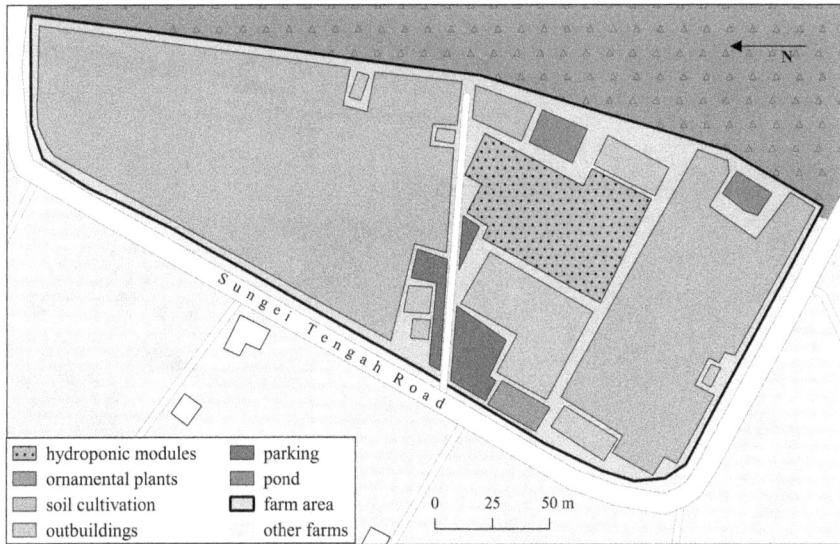

Figure 4.20 A diagram of spatial development of the parcel occupied by *Kok Fah Hydro-ponic Farms*, located within the limits of Kranji Countryside.

Source: Own study on the basis of field research.

4.6.4 *Kok Fah Technology Farm*

The second farm located in Kranji Countryside and described in detail is *Kok Fah Technology Farm*, situated in Sungei Tengah Road at the fork of the rivers Tengah and Peng Siang. The area exhibits a high density of urban farms, both those specialising in production of edible plants and those involved in growing decorative plants and aquaculture.

Kok Fah Technology Farm was established 1979 and is an example of a farm which, due to its location as well as high profitability, will most likely remain in Singapore for a longer time. It is situated in an area allocated to urban agriculture under the current Master Plan. The farm covers 3.67 ha, of which nearly 2.2 ha is taken up by soil cultivation and 0.3 ha by hydroponic cultivation in a roofed facility (see Figure 4.20). *Kok Fah* also features eight farm buildings, tents with decorative plants intended for sale and a small retention tank. Along the internal road, there are also three small parking lots for delivery vehicles.

The farm specialised in the production of Asian leafy greens (kai xin, kai lan, kang kong and spinach), herbs (e.g. mint), as well as different varieties of lettuce. The majority of products come from semi-protected soil cultivation, and it is those products that are intended for sale to Singaporean supermarkets and at the same time form the core of the company's activity (see Figure 4.10).

According to the respondent, the hydroponic modules located in a room with strictly controlled temperature, lighting and humidity were exclusively experimental and demonstrative (see Figure 4.21). Vegetables grown using this method could

Figure 4.21 Peppermint cultivation in A-frame hydroponic modules.
Source: Photograph by K. Górny, January 2019.

be purchased on site. The number of modules and thus the production yield was not big enough to supply larger stores. However, the respondent emphasised that conducted research and expert analyses demonstrated that hydroponics is an efficient production method. That is why the farm owners have decided to purchase more land in order to develop it further. Moreover, a part of the demonstration/experimental room is also dedicated to growing cherry tomatoes in pots filled with coconut fibres. According to the respondent, all seeds the farm used were imported from Europe (among others, lettuce, tomatoes) and China (Asian leafy greens). According to information published on the website, the farm also has a seed nursery.

The farm uses compost produced on site as a substitute for artificial fertilisers, while pesticides are only used in the case of plant diseases. When their use is not necessary, protection is provided by flypaper placed in the immediate vicinity of plant beds and protective nets (see Figure 4.10). This mechanical method is only applied together with soil cultivation.

Kok Fah employs 80 people – the largest number among all farms analysed in Singapore. The employees are involved not only in the production process, but also in conducting trainings and workshops as well as organising educational tours, which serve as additional income-generating and promotional activity of the farm. According to the respondent, *Kok Fah* is typically visited by school trips from educational institutions all over Singapore. Foreign tourists are less frequent visitors.

Apart from organising training and educational events, the farm has expanded its activity to include construction and installation of hydroponic systems in other locations in the city. Thanks to financial support of national institutions provided to the company in the form of grants, *Kok Fah* is currently capable of developing, diversifying its sources of income and investing in modern technologies. The farm's model of operation can be considered the most desirable for the authorities of Singapore, thanks to which the future of *Kok Fah* is not at risk.

4.7   Urban agriculture in Singapore – a summary

In view of the direction of development of agriculture in Singapore, clearly outlined in the authorities' policy, Singaporean farms seem to be experiencing an important transitional moment. On the one hand, some facilities, in the spirit of change imposed by the government, are adapting their activities to the vision of modern Singaporean agriculture, which fits in with the smart city concept. They are introducing modern technological solutions, applying increasingly intensive production methods that ensure high profitability of their ventures and striving to include urban agriculture in an innovative model of conducting business activity. Singapore can be therefore considered somewhat of a laboratory of modern urban agriculture and the technological solutions used in the city can be successfully implemented in other metropolises worldwide. On the other hand, some farms in Singapore still apply conventional cultivation techniques, emphasising their social and educational functions. Their activity is not profitable enough for them to compete on the market with the representatives of the former group. Urban agriculture in Singapore is therefore at a turning point, where the technocratic vision of a city of the future is starting to prevail over grassroots initiatives and needs of the residents as well as over tradition, while low-profit forms of agricultural activities are being removed from the city space. Investments in cutting-edge technologies and drastic policy towards insufficiently efficient farms are the Singaporean government's method of reaching the goal established in March 2019, which was achieving a 30% level of food self-sufficiency by 2030, also referred to as the "30 by 30" goal (Singapore Food Agency 2021a), while simultaneously taking over and developing land previously allocated to traditional crops. However, there is no doubt that the authorities' activities are leading to restriction of grassroots initiatives of the residents, which may be less technologically advanced but strengthen social bonds. They also educate the school-age youth typically deprived of contact with the countryside and unfamiliar with the methods of production of goods consumed on a daily basis. Moreover, the government's decisions cause discontent among residents who have been investing in their agricultural activity for years in order to ultimately be forced to give up on it, as demonstrated by the interviews conducted with representatives of urban farms.

Singaporean authorities have announced widespread reforms aimed at implementing the assumptions of the smart city concept; however, they are only focusing on technocratic solutions and options that generate the highest possible profits. It should be emphasised that the concept itself has evolved over the last three decades, and currently, cities considered smart cities are those in which the residents are the main decision-makers with real impact on the pursued policy, whose needs are at the centre of attention. Thus, the actions of Singaporean authorities only fit in with the assumptions of the smart city idea selectively since they omit an incredibly important actor – the residents. Stiff institutional and legal framework prevents the residents from taking action that does not fit in with the authorities' policy. This hampers grassroots initiatives and limits the residents' participation in shaping urban space, leading to a situation where they become merely its users, not free creators.

Notes

1 Thirty-seven farms were visited during the field research. However, one of them – *Goodland Hydroponic Farm* – located in the northern part of the city, turned out to be shut down, so it was excluded from further analysis.
2 This sub-chapter, as well as sub-chapters 4.2–4.6, uses fragments of an original paper: Górna, A., & Górny, K. (2021). Singapore vs. the 'Singapore of Africa'—Different Approaches to Managing Urban Agriculture. *Land*, *10*(9), 987, https://doi.org/10.3390/land10090987 (Creative Commons CC BY 4.0 Licence).
3 The names in Javanese and Malay mean "the sea".
4 The Ministry of Environment has now been transformed into the Ministry of Sustainability and Environment, which clearly shows the authorities' approach to issues related to environmental protection.
5 Detailed information was obtained during a semi-structured interview conducted in February 2019 with an employee of *Hay Dairies*. Moreover, the information was supplemented with content available on the farm's official website (Hay Dairies, 2021).
6 This information was obtained from a representative of the SFA – Assistant Director of the Agri-Tech & Food Innovation Department.

References

Abshire, J. (2011). *The history of Singapore. The Greenwood Histories of the Modern Nations*. Greenwood: Bloomsbury Academic.

Albino, V., Berardi, U., & Dangelico, R. M. (2015). Smart cities: Definitions, dimensions, performance, and initiatives. *Journal of Urban Technology*, *22*(1), 3–21, https://doi.org/10.1080/10630732.2014.942092

Anas, A., Arnott, R., & Small, K. A. (1998). Urban spatial structure. *Journal of Economic Literature*, *36*(3), 1426–1464, https://www.jstor.org/stable/2564805

Astee, L. Y., & Kishnani, N. T. (2010). Building integrated agriculture: Utilising rooftops for sustainable food crop cultivation in Singapore. *Journal of Green Building*, *5*(2), 105–113, https://doi.org/10.3992/jgb.5.2.105

Benke, K., & Tomkins, B. (2017). Future food-production systems: Vertical farming and controlled-environment agriculture. *Sustainability: Science, Practice and Policy*, *13*(1), 13–26, https://doi.org/10.1080/15487733.2017.1394054

Chang, T. C., & Huang, S. (2011). Reclaiming the city: Waterfront development in Singapore. *Urban Studies*, *48*(10), 2085–2100, https://doi.org/10.1177/0042098010382677

Channel New Asia. (2019). *MPs question Government's stand on single-use plastics as Parliament passes new sustainability Bill*, https://www.channelnewsasia.com/news/singapore/singapore-single-use-plastics-resource-sustainability-bill-11872642 (accessed 31.01.2020).

Chee, S. Y., Othman, A. G., Sim, Y. K., Adam, A. N. M., & Firth, L. B. (2017). Land reclamation and artificial islands: Walking the tightrope between development and conservation. *Global Ecology and Conservation*, *12*, 80–95, https://doi.org/10.1016/j.gecco.2017.08.005

Cheng, C. S. (1995). *The Singapore improvement trust and pre-war housing.* ScholarBank@ NUS Repository, https://scholarbank.nus.edu.sg/handle/10635/153078

Cheong, C. C., & Toh, R. (2010). Household interview surveys from 1997 to 2008. A decade of changing travel behaviours. *Journeys, 4,* 52–61.

Chew, V. (2009). *History of urban planning in Singapore.* National Library Board, https://eresources.nlb.gov.sg/infopedia/articles/SIP_1564_2009-09-08.html

Cohen, B. (2014). *The smartest cities in the world 2015,* https://www.fastcompany.com/3038818/the-smartest-cities-in-the-world-2015-methodology

Comcrop. (2020). *Singapore's first and only commercial rooftop farming company,* http://comcrop.com/ (accessed 22.11.2020).

Cornelius, V. (1999). *Orchard Road,* https://web.archive.org/web/20131029202140/http://infopedia.nl.sg/articles/SIP_721_2005-01-03.html (accessed 25.06.2023).

Davies, N. (1998). *Europe: A history,* HarperCollins Publishers, Inc, https://doi.org/10.2307/494453

Davies, N. (2017). *Beneath another sky: A global journey into history.* Penguin UK.

Discovery Channel. (2016). *Growing roots – this farmer is taking root on your rooftops,* https://www.youtube.com/watch?v=vPRySy3Qtvs&t=194s&ab_channel=Viddsee (accessed 22.11.2020).

Diver, S., & Rinehart, L. (2000). Aquaponics-Integration of hydroponics with aquaculture. *Attra,* 1–16.

Dos Santos, M. J. P. L. (2016). Smart cities and urban areas. Aquaponics as innovative urban agriculture. *Urban Forestry & Urban Greening, 20,* 402–406, https://doi.org/10.1016/j.ufug.2016.10.004

Eng, T. S. (1986). New towns planning and development in Singapore. *Third World Planning Review, 8*(3), 251, https://doi.org/10.3828/twpr.8.3.fm13m66575663j07

Gibson, D. V., Kozmetsky, G., & Smilor, R. W. (1992). *The technopolis phenomenon: Smart cities, fast systems, global networks.* Rowman & Littlefield, https://doi.org/10.1002/bs.3830380207

Giffinger, R., Fertner, C., Kramar, H., Kalasek, R., Pichler-Milanovic, N., & Meijers, E. J. (2007). *Smart cities: Ranking of European medium-sized cities.* Vienna University of Technology, https://doi.org/10.34726/3565

Górna, A., & Górny, K. (2019). Perspektywy wykorzystania rolnictwa miejskiego jako nowej formy turystyki miejskiej – przykład Singapuru [Prospects for the use of urban agriculture as a new form of urban tourism – the example of Singapore]. *Turystyka Kulturowa, 2,* 69–83.

Górna, A., & Górny, K. (2021). Singapore vs. the 'Singapore of Africa'—different approaches to managing urban agriculture. *Land, 10*(9), 987, https://doi.org/10.3390/land10090987

Govada, S. S., Spruijt, W., & Rodgers, T. (2017). Smart city concept and framework. In *Smart economy in smart cities. Advances in 21st century human settlements,* ed. Vinod Kumar, T. Singapore: Springer, https://doi.org/10.1007/978-981-10-1610-3_7

Government of Singapore. (2019). *The Planning Act Master Plan Written Statement 2019.*

Haila, A. (2015). *Urban land rent: Singapore as a property state.* John Wiley & Sons.

Han, S. S. (2005). Global city making in Singapore: A real estate perspective. *Progress in Planning, 2*(64), 69–175, https://doi.org/10.1016/j.progress.2005.01.001

Harvey, D. (2008). The right to the city. *The City Reader, 6*(1), 23–40.

Hay Dairies. (2021). *Farm fresh goodness,* https://haydairies.sg/ (accessed 15.01.2021).

Henderson, J. C. (2009). Agro-tourism in unlikely destinations: A study of Singapore. *Managing Leisure, 14*(4), 258–268, https://doi.org/10.1080/13606710903204456

Henderson, J. C. (2012). Planning for success: Singapore, the model city-state? *Journal of International Affairs*, *65*(2), 69–83, https://www.jstor.org/stable/24388219

Henderson, J. C. (2013). Urban parks and green spaces in Singapore. *Managing Leisure*, *18*(3), 213–225, https://doi.org/10.1080/13606719.2013.796181

Ho, E. (2017). Smart subjects for a Smart Nation? Governing (smart) mentalities in Singapore. *Urban Studies*, *54*(13), 3101–3118, https://doi.org/10.1177/004209801666430

Howard, E. (1902). *Garden cities of to-morrow*. London: Swan Sonnenschein & Co.

IMD. (2019). *Smart City Index 2019*, https://www.imd.org/research-knowledge/reports/imd-smart-city-index-2019/ (accessed 22.10.2019).

IMD. (2020). *Smart City Index 2020*, https://imd.cld.bz/Smart-City-Index-2020 (accessed 19.06.2023).

IMD. (2021). *Smart City Index 2021*, https://imd.cld.bz/Smart-City-Index-2021 (accessed 19.06.2023).

IMD. (2023). *Smart City Index 2023*, https://www.imd.org/smart-city-observatory/home/ (accessed 19.06.2023).

Jensen, M. H. (1997). Hydroponics. *HortScience*, *32*(6), 1018–1021, https://doi.org/10.21273/HORTSCI.32.6.1018

Kao, T. C., Hsiang, T., & Changhua, R. O. C. (1991). *The dynamic root floating hydroponic technique: Year-round production of vegetables in roc on Taiwan*. ASPAC Food & Fertilizer Technology Center, 1–17.

Kong, L., & Yeoh, B. S. (1994). Urban conservation in Singapore: A survey of state policies and popular attitudes. *Urban Studies*, *31*(2), 247–265, https://www.jstor.org/stable/24388219

Kong, L., & Yeoh, B. S. (1996). Social constructions of nature in urban Singapore. *Japanese Journal of Southeast Asian Studies*, *34*(2), 402–423, https://doi.org/10.20495/tak.34.2_402

Kumar, T. V., & Dahiya, B. (2017). Smart economy in smart cities. In *Smart economy in smart cities*, ed. T. M. Vinod Kumar. Singapore: Springer, 3–76, https://doi.org/10.1007/978-981-10-1610-3_1

Lai, S., Loke, L. H., Hilton, M. J., Bouma, T. J., & Todd, P. A. (2015). The effects of urbanisation on coastal habitats and the potential for ecological engineering: A Singapore case study. *Ocean & Coastal Management*, *103*, 78–85, https://doi.org/10.1016/j.ocecoaman.2014.11.006

Leitch Lepoer, B. (1989). *Singapore: A country study*. Washington: GPO for the Library of Congress, http://countrystudies.us/singapore/39.htm (accessed 6.03.2021).

Leitmann, J. (2000). *Integrating the environment in urban development: Singapore as a model of good practice*. Urban Development Division, World Bank, Washington, 1–24.

Mahizhnan, A. (1999). Smart cities: The Singapore case. *Cities*, *16*(1), 13–18, https://doi.org/10.1016/S0264-2751(98)00050-X

Maye, D. (2019). 'Smart food city': Conceptual relations between smart city planning, urban food systems and innovation theory. *City, Culture and Society*, *16*, 18–24, https://doi.org/10.1016/j.ccs.2017.12.001

Monfaredzadeh, T., & Berardi, U. (2015). Beneath the smart city: Dichotomy between sustainability and competitiveness. *International Journal of Sustainable Building Technology and Urban Development*, *6*(3), 140–156, https://doi.org/10.1080/2093761X.2015.1057875

Motha, P., & Yuen, B. (1999). *Singapore real property guide*. Singapore: Singapore University Press.

National Environment Agency. (2020). *Waste statistics and overall recycling*, https://www.nea.gov.sg/our-services/waste-management/waste-statistics-and-overall-recycling (accessed 28.02.2020).

Neo, W. A. (2022). *A state to house a nation: The Singapore Improvement Trust, 1947-1959.* ScholarBank@NUS Repository, https://scholarbank.nus.edu.sg/handle/10635/228492

Newman, P. (2014). Biophilic urbanism: A case study on Singapore. *Australian Planner, 51*(1), 47–65, https://doi.org/10.1080/07293682.2013.790832

Nyerere, J. (2000). Foreword. In *What is Africa's problem?*, ed. Y. Museveni. Minneapolis: University of Minnesota Press.

Qonita, M., & Giyarsih, S. R. (2023). Smart city assessment using the Boyd Cohen smart city wheel in Salatiga, Indonesia. *GeoJournal, 88*(1), 479–492, https://doi.org/10.1007/s10708-022-10614-7

Roth, M., & Chow, W. T. (2012). A historical review and assessment of urban heat island research in Singapore. *Singapore Journal of Tropical Geography, 33*(3), 381–397, https://doi.org/10.1111/sjtg.12003

Rowe, P. G., & Hee, L. (2019). *A city in blue and green: The Singapore story.* Springer Nature, https://doi.org/10.1007/978-981-13-9597-0

Santhi, S., & Saravanakumar, A. R. (2020). The economic development of Singapore: A historical perspective. *Aut Aut Research Journal, 11*(7), 441–459.

Savage, V. R., & Kong, L. (1993). Urban constraints, political imperatives: Environmental 'design' in Singapore. *Landscape and Urban Planning, 25*(1–2), 37–52, https://doi.org/10.1016/j.cities.2013.02.001

Sen, A. (2001). *Development as freedom.* Oxford Paperbacks.

Shah, M. N., Nagargoje, S., & Shah, C. (2017). Assessment of Ahmedabad (India) and Shanghai (China) on smart city parameters applying the Boyd Cohen smart city wheel. In *Proceedings of the 20th international symposium on advancement of construction management and real estate.* Singapore: Springer, 111–127, https://doi.org/10.1007/978-981-10-0855-9_10

Shamshiri, R., Kalantari, F., Ting, K. C., Thorp, K. R., Hameed, I. A., Weltzien, C., & Shad, Z. M. (2018). Advances in greenhouse automation and controlled environment agriculture: A transition to plant factories and urban agriculture. *International Journal of Agriculture & Biological Engineering, 11*(1), 1–22, https://doi.org/10.25165/j.ijabe.20181101.3210

Shatkin, G. (2014). Reinterpreting the meaning of the 'Singapore Model': State capitalism and urban planning. *International Journal of Urban and Regional Research, 38*(1), 116–137, https://doi.org/10.1111/1468-2427.12095

Singapore Department of Statistics. (2023). *Environment,* https://www.singstat.gov.sg/publications/reference/ebook/society/environment (accessed 25.06.2023).

Singapore Food Agency. (2020a). *Food farms – farm employment,* https://www.sfa.gov.sg/food-farming/food-farms/farm-employment (accessed 12.11.2020).

Singapore Food Agency. (2020b). *Food farms,* https://www.sfa.gov.sg/food-farming/food-farms (accessed 9.10.2020).

Singapore Food Agency. (2020c). *Starting a farm,* https://www.sfa.gov.sg/food-farming/food-farms/starting-a-farm (accessed 3.10.2020).

Singapore Food Agency. (2021a). *Food farming,* https://www.sfa.gov.sg/food-farming (accessed dostęp 17.01.2021).

Singapore Food Agency. (2021b). *The food we eat,* https://www.sfa.gov.sg/food-farming/singapore-food-supply/the-food-we-eat (accessed 12.01.2021).

Singapore Government Agency. (1991). *1991 Concept Plan is unveiled.* HistorySG. An online resource guide, https://eresources.nlb.gov.sg/history/events/8a9c774e-3e4a-46c8-8862-0ff9f4ae257e (accessed 9.01.2021).

Singapore Land Authority. (2023). https://www.sla.gov.sg/properties/land-sales-and-lease-management/lease-management/lease-policy (accessed 27.11.2023).

Smart Nation and Digital Government Office. (2023). *Smart Nation Digital Government Group,* https://www.smartnation.gov.sg/about-smart-nation/sndgg/ (accessed 25.06.2023).

Solarz, M. W. (2014). *The language of global development: A misleading geography*. Rout-ledge, https://doi.org/10.4324/9780203077382

Tan, K. W. (2006). A greenway network for Singapore. *Landscape and Urban Planning*, *76*(1-4), 45–66, https://doi.org/10.1016/j.landurbplan.2004.09.040

Tan, L. H., & Neo, H. (2009). "Community in Bloom": Local participation of community gardens in urban Singapore. *Local Environment*, *14*(6), 529–539, https://doi.org/10.1080/13549830902904060

Tan, P. Y., Wang, J., & Sia, A. (2013). Perspectives on five decades of the urban greening of Singapore. *Cities*, *32*, 24–32, https://doi.org/10.1016/j.cities.2013.02.001

Tay, T. F. (2016). Agriculture land leases to be doubled to 20 years; 62 Lim Chu Kang farms will have tenures extended to end-2019. *The Straits Times*, https://www.straitstimes.com/singapore/agriculture-land-leases-to-be-doubled-to-20-years-ava (accessed 25.06.2023).

Tey, Y. S., Suryani, D., Emmy, F. A., & Illisriyani, I. (2009). Food consumption and expen-ditures in Singapore: Implications to Malaysia's agricultural exports. *International Food Research Journal*, *16*(2), 119–126.

The Straits Times. (2019). *Parliament: New zero-waste law to compel big firms to take greater action*, https://www.straitstimes.com/singapore/new-zero-waste-law-to-compel-big-firms-to-take-greater-action (accessed 31.01.2020).

Thompson, H. C., Langhans, R. W., Both, A. J., & Albright, L. D. (1998). Shoot and root temperature effects on lettuce growth in a floating hydroponic system. *Journal of the American Society for Horticultural Science*, *123*(3), 361–364, https://doi.org/10.21273/JASHS.123.3.361

Turnbull, C. M. (2009). *A history of modern Singapore, 1819-2005*. Singapore: Nus Press.

United Nations Development Programme. (2022). *Human Development Report 2021/2022: Uncertain times, unsettled lives: Shaping our future in a transforming world*, https://reliefweb.int/report/world/human-development-report-20212022-uncertain-times-unsettled-lives-shaping-our-future-transforming-world-enruzh?gclid=Cj0KCQjwy9-kBhCHARI-sAHpBjHhDQEBgB6QqT2rTx7ai96r13OdquKz1b-hM9d7m7d06djSgtMKiV0oaAoN5EALw_wcB

Whitehead, R. (2019). Singapore farmers hope government gives clarity on the future of their lands. *Dairy Reporter*, https://www.dairyreporter.com/Article/2019/03/27/Singapore-farmers-hope-government-gives-clarity-on-the-future-of-their-lands (accessed 2.10.2020).

Woo, J. J. (2017). *Singapore's smart nation initiative – a policy and organisational perspec-tive*. Lee Kuan Yew School of Public Policy, National University of Singapore.

Yuen, B. (2009). Guiding spatial changes: Singapore urban planning. *Urban Land Markets: Improving Land Management for Successful Urbanization*, 363–384, https://doi.org/10.1007/978-1-4020-8862-9_14

Zhong, C., Arisona, S. M., Huang, X., Batty, M., & Schmitt, G. (2014). Detecting the dy-namics of urban structure through spatial network analysis. *International Journal of Geographical Information Science*, *28*(11), 2178–2199, https://doi.org/10.1080/1365 8816.2014.914521

Zhong, C., Huang, X., Arisona, S. M., & Schmitt, G. (2013). *Identifying spatial structure of urban functional centers using travel survey data: A case study of Singapore*, 28–33, https://doi.org/10.1145/2534848.2534855

5 Kigali – a case study of urban agriculture in Sub-Saharan Africa

The last case study analysed in this book is Kigali – the capital city of Rwanda, with a population of more than one million people. Due to its dynamic development, reflected, among others, in the investments in the services sector and eco-friendly solutions implemented by the authorities, Rwanda – and therefore its capital – is frequently referred to as "the Singapore of Africa" in media discourse (*The Economist Newspaper* 2012). Thanks to a high share of greenery within the city space and, in particular, clean streets that distinguish it among other metropolises on the continent, Kigali is indeed reminiscent of the Asian city-state discussed in the previous chapter. What is also important is the fact that Rwanda's authorities have been following the example of the Singaporean experience in spatial planning for years. Lee Kuan Yew, the co-founder of the People's Action Party and a long-serving prime minister of Singapore (1959–1990), advised the President of Rwanda, Paul Kagame, on efficient country management during the first decade of the 21st century (Goodfellow 2014). What is more, Kigali is one of few examples of African metropolises where the premises of the smart city concept are being implemented, similar to Singapore. The Smart Kigali initiative is an example of measures which fit within the framework of this idea. Its objectives are, among others, to give the residents access to a free Wi-Fi connection as well as inexpensive smartphones (Masłoń-Oracz & Mazurewicz 2015).

However, in terms of the characteristics of urban agriculture, the Rwandan capital does not resemble Singapore. In that regard, it is representative of Sub-Saharan Africa. In Kigali, similar to Bissau, Brazzaville, Yaoundé or Kampala, large-area agricultural zones are concentrated at the bottom of the river valleys and are arranged in a wedge-ribbon layout, playing a prominent role in the city's spatial and functional structure (Górna & Górny 2020). Therefore, on the one hand, Rwanda's capital is an example of a dynamically developing city, following in Singapore's tracks when it comes to taking the first steps towards developing a sustainable, modern urban system; on the other hand, it is a typical representative of African cities in terms of the characteristics and high share of agriculture in the urban space.

Based on an analysis of satellite images available in Google Earth Pro,[1] a densely developed area of 9,180 ha in total was distinguished in Kigali, with 780 agricultural zones identified within its borders. Ninety-eight[2] of them were subjected to a detailed field analysis conducted in July 2019. Steps were taken to ensure that the

DOI: 10.4324/9781003429845-5

selected sample included areas with various direct characteristic features, including shape, size, structure, colour on the orthophotomap (which may indicate the diversity of crop types) as well as various indirect characteristics, such as the shade cast by facilities within individual areas, the location and associations with other landscape features. The purpose of it was to ensure that the analysed sites were as representative as possible. Moreover, when selecting individual areas, their accessibility was also taken into account since it determined the success of the fieldwork. For example, analysis of small agricultural zones located between private buildings, with particularly impeded access, was omitted. On the other hand, the analysed sample included many easily accessible large-area agricultural zones situated at the bottom of the valleys.

In all cases covered by the field research, the presence of urban agriculture identified on the orthophotomap during the first stage of the analysis was confirmed, which demonstrates the effectiveness of the selected method of manual and visual interpretation of satellite and aerial images. The field research was carried out according to previously designated routes. The 98 visited agricultural zones underwent detailed field observations in accordance with the developed electronic questionnaire form. It covered both internal characteristics of the areas themselves and the features of their surroundings. Another important element of the fieldwork was collection of photographic documentation serving as the basis for preparing cartographic studies, including cross-sections of the valley bottoms used for crop cultivation.

The specificity of the city discussed in this chapter prevented a satisfactory number of semi-structured interviews from being conducted with the food producers. This was due to a number of previously unforeseen difficulties which arose during the field research. First of all, large-area agricultural zones in Kigali, located at the bottom of the vast valleys, are in all cases divided into dozens of smaller fields leased by individual residents. This prevented identification of the people in charge of individual sections of the parcels, and thus interviews with them. Second, the majority of fields on hill slopes were located in fenced-off private areas, as a result of which reaching potential respondents proved unsuccessful. Another significant impediment was the fact that the producers with whom interviews were attempted spoke neither English nor French despite both languages being official languages in Rwanda. The majority of the population engaged in urban agriculture speak local languages. In this case, it was mainly Kinyarwanda and less frequently Kiswahili, which is common in East Africa. Therefore, interviews with producers proved impossible without an interpreter, whose services had not been taken into account for financial reasons. Despite the outlined difficulties, three semi-structured interviews were successfully conducted during the field research – with a woman working at one of the crop fields (in the presence of a spontaneously encountered person who spoke both Kinyarwanda and English) and two men working at local crop sales points (also in English). An important addition to the field observations was an in-depth interview with a representative of the Kigali City Hall in charge of implementation of the Kigali Master Plan published in 2013. The interview was conducted at the city hall and pertained to the presence and function of agriculture in

Kigali as well as the scope of protection of the cultivated crops under the Rwandan strategic and planning documents. The interview enabled a broader understanding of the role played by urban agriculture within the spatial and functional structure of the city as well as provided information on the authorities' policy directions within the framework of its development prospects.

The first part of this chapter is dedicated to Kigali's spatial and functional structure as well as its development from the colonial times until present day, with particular emphasis on a number of pro-environmental reforms implemented during President Paul Kagame's term in office. This is followed by a description of the institutional and legal framework, including primarily the strategic planning documents which shape the development of urban agriculture in Kigali. The next sub-chapter, based on the results of stationary and field research, will take into account the following issues: distribution of urban agriculture in Rwanda's capital city, its characteristics and the functions it fulfils in the city's spatial and functional structure (similar to the analogous sub-chapters dedicated to Havana and Singapore).

5.1   Spatial and functional structure of the city[3]

Rwanda is a small inland country situated in East Africa. It borders Uganda from the north, Burundi from the south, Tanzania from the east and the Democratic Republic of the Congo from the west. Kigali is the capital and, at the same time, the largest city of Rwanda, located 215 km to the south of the equator. It occupies several dozen hills separated by a number of vast river valleys. As a result, it has particularly diverse terrain. The height differences reach 450 m, and the highest peak in the area is Mount Kigali (1,850 m above sea level) located in the south-west part of the city. The elevation incline ranges from 2% to as much as 50% (Manirakiza et al. 2019). According to official data published by the city hall, 35% of the city area has an incline of more than 20%, 14% of which are developed areas (City of Kigali 2013). Construction of informal residential buildings on steep slopes increases erosion and contributes to a higher risk of mass wasting (primarily landslides), threatening the safety of the residents (ibidem). In 2020, more than 19% of the city's area was deemed unsuitable for development precisely because of the excessive incline (over 30%) (City of Kigali 2020). Apart from the diverse terrain, another prominent feature of the city's natural environment is a dense network of watercourses and canals. The largest river in Kigali is the Nyabugogo, flowing towards the south-west and into the Nyabarongo on the border between the Kigali Province and the Southern Province. Other larger watercourses within the administrative borders of Kigali include Yanze, Kibumba, Rwazangoro and Ruganwa. Due to the varied terrain, there are as many as 25 water divides within the city, separating the river basins of Lake Victoria's catchment area (Manirakiza et al. 2019). Wetlands are an important feature of Kigali's landscape. They currently cover approximately 12.5% of the city's total surface area. The capital of Rwanda is located in the tropical savannah climate zone (according to the Köppen climate classification), with a short dry season lasing from June to August. Due to the high altitude, the average annual temperature is approximately 20°C, while the amplitude between the temperatures during the dry and the wet seasons is not significant.

In administrative terms, Kigali is one of five provinces of the country established in 2006, in addition to the Northern, Western, Eastern and Southern Provinces. The city is divided into three districts – *Gasabo, Kicukiro* and *Nyarugenge*. Those, in turn, are divided into 35 smaller sectors – *Gasabo* into 15, while *Kicukiro* and *Nyarugenge* into ten each. The area of the city is 730 km², of which only 17% is taken up by urbanised zones (City of Kigali 2013). The remaining 83% is occupied by green areas and agricultural land, typical of rural areas. The historical core and the contemporary functional centre of the city are primarily concentrated on the *Nyarugenge* hill and in areas stretching towards the east, all the way to the Kigali International Airport. In 2019, the city's population amounted to more than one million – almost 9% of the entire population of Rwanda. The population density varies depending on the district and amounts to 2,127 people/km² in *Nyarugenge*, 1,918 people/km² in *Kicukiro* and 1,237 people/km² in *Gasabo* (Benken 2017).

When analysing the spatial and functional structure of Kigali, it is crucial to study the transformation that the Rwandan capital city has undergone: from a minor colonial settlement to a city considered a "model" example of spatial planning in Sub-Saharan Africa (Goodfellow & Smith 2013; Goodfellow 2014). This subchapter will first describe the original traces of settlement in the area of modern-day Kigali, then the city's development during the colonial period and right after the country gained independence. This will be followed by a presentation of the changes which have taken place in the Rwandan capital in the aftermath of the 1994 genocide, in both spatial and demographic terms. Finally, the pro-environmental reforms implemented by the current authorities of Rwanda will be portrayed along with contemporary challenges within the scope of sustainable and inclusive city development.

5.1.1   The pre-colonial period

As far as sources on pre-colonial history are concerned, Rwanda shares the fate of many countries of Central Africa. Their historiography is extremely limited due to the complete lack or considerable shortage of written sources preceding the colonial period. In Rwanda's case, the first written texts date back to late 19th century. They are documents drawn up by European explorers, missionaries, colonial officers, state officials and entrepreneurs (Byanafashe & Rutaysire 2016). Sources of information on earlier times are mainly oral, passed down from generation to generation in the form of spoken tales (Rennie 1972), as well as derived from the results of archaeological works (Byanafashe & Rutaysire 2016). However, it should be emphasised that the quality of those sources is dubious and archaeological sites in the area of modern-day Rwanda have either not been identified so far or are insufficiently protected, hence the findings obtained there can hardly be considered abundant (ibidem). The territory of modern-day Kigali is not mentioned in written sources until the founding of the city in 1907 by European settlers. The area of the modern-day *Gasabo* district, and more precisely the hill itself, is considered to be the cradle of the Kingdom of Rwanda, which by the 20th century had expanded its influence onto the neighbouring settlements located on the surrounding hills – *Buganza, Buriza, Busigi* and *Burasi* (Byanafashe & Rutaysire 2016).

5.1.2   *The colonial period*

The first Europeans arrived in Rwanda in the mid-19th century. Their presence was associated with the search for the sources of the Nile, commenced in 1856 by the London Geographical Society. An important figure in the history of Kigali, also associated with widespread search for the sources of the longest river in Africa, was an explorer of Polish origin, Richard Kandt (born in 1867 in German-occupied Poznań as Ryszard Kantorowicz). Along with his companion – Jan Czekanowski, an anthropologist – he led one of the first ethnographic and anthropological research initiatives concentrated around Lake Kivu (ibidem).

As a result of the arrangements made during the Berlin Conference (1884–1885), the area of modern-day Rwanda was given to Germans, who officially incorporated it into the German East Africa in 1890 by merging the areas of Tanganyika (the continental part of modern-day Tanzania), Burundi and Rwanda into a single administrative body. Richard Kandt, by virtue of his previous academic achievements as well as knowledge of the local context, was appointed the first German resident in Rwanda and chose the *Nyarugenge* hill as his place of residence.[4] Kigali was founded in 1907 on that very hill as the capital city of Rwanda, which was separated administratively the same year from Burundi, located to the south of it (ibidem). Thanks to its strategic location, in the first years of its existence, what was at the time a small settlement quickly became a communications hub for transport of goods between Bukoba and Kigoma in Tanganyika and Bujumbura in Burundi, as well as between Kisangani in the Belgian Congo and Kampala in the British Uganda (Bindseil 1988, as quoted in Manirakiza et al. 2019). Kigali was the reloading point for ivory, leathers and furs on the trading route leading in particular to Bukoba on Lake Victoria, from where the goods obtained inland were transported further to European-controlled ports on the coast of the Indian Ocean (Byanafashe & Rutaysire 2016). The prestigious position on the regional market for trade in goods attracted many Arab and Indian settlers (Manirakiza et al. 2019). Germans ruled over Rwanda only for a brief period of time, until 1916. During World War I, its territory was occupied by the troops of Belgium, a supporter of the Allies, which was in charge of the neighbouring Congo. In 1923, Rwanda officially became a part of a territory with the League of Nations' mandate – Ruanda-Urundi, which was administered by Belgians (Záhořík 2011; De Haas 2019).

During the first period of its existence, the settlement established by Germans and taken over by Belgians during World War I played a marginal role in the system of European colonial administration. Compared with other cities of East Africa, such as the ports of Dar es Salaam and Mombasa or inland cities of Nairobi and Kampala, Kigali was a small trading town. Before Rwanda gained independence, its capital city was only a small village that primarily fulfilled administrative functions (Sirven 1884, as quoted in Manirakiza 2014, p. 506). Kigali was dominated by low-rise informal buildings and the majority of its area was occupied by marginal districts (Goodfellow & Smith 2013). Towards the end of colonial domination, the city was characterised by a slow pace of demographic and spatial development (Manirakiza 2014). On the eve of independence, gained by Rwanda on 1 July 1962, Kigali had merely 6,000 residents and occupied only 2.5 km² (Manirakiza 2014; Nduwayezu et al. 2016; Manirakiza et al. 2019).[5] Since its founding in 1907, it had not expanded spatially beyond the *Nyarugenge* hill (Manirakiza 2014).

5.1.3   The first decades of independence

The situation began to change slowly after 1962, when the city became the capital of an independent country. It was then that Kigali experienced a dynamic population increase, followed by intensified spatial expansion. To control the intensive development, the first Kigali Master Plan was already implemented in 1964 (Conceptual Master Plan for Kigali) (Nduwayezu 2015). It featured four main postulates, whose fulfilment was supposed to improve the living conditions of the residents and organise the city's spatial structure. The first one was expanding the administrative borders of Kigali to 70 km², the second one was making the slopes less steep so that they could be built over, the third one was the construction of the French ministry offices,[6] while the fourth one was setting out inexpensive parcels for residential development in *Nyamirambo*,[7] located to the south of the *Nyarugenge* hill (Manirakiza 2014; Benken 2017). The last measure was meant to funnel the new residents into less densely populated areas in order to take the stress off of the densely developed core of the city. The first two objectives were already achieved in 1969. However, construction of French ministry offices proved impossible in the originally selected location. The offices of the French Ministry of Cooperation were built on the *Kacyiru* hill (to the north-east of *Nyarugenge*), which is nowadays also occupied by the headquarters of international organisations and foreign diplomatic missions. Delimitation of housing estates in *Nyamirambo*, on the other hand, proved impossible due to the extremely fast increase in the number of informal buildings (Benken 2017).

Already in 1975, Kigali extended beyond the spatial borders set out in 1962. Nearly a decade later, in 1984, the city's population stood at 160,000 (i.e. nearly 27 times more than when the country gained independence), while the urbanised area had increased sixfold to as much as 15 km² (Niyonsenga 2013, as quoted in Benken 2017). In the early 1980s, the French Ministry of Cooperation, financing the majority of investments in the city, once again pressured the authorities of Kigali to implement a new Master Plan that assumed construction of inexpensive housing estates in order to limit informal development. This time, the housing estates would be located not only in *Nyamirambo*, but also on the *Gikondo, Kicukiro, Remera, Kimihurura* and *Kacyiru* hills. The assumptions of the plan were only partially implemented. Although new homes were eventually constructed, the current housing estates located in *Kicukiro, Kimihurura* and *Kacyiru* are mainly inhabited by Kigali elites. They are expensive and unavailable to the city's poorest population, which contradicts the planners' original objectives. The Master Plan introduced in the 1980s also assumed development of industrial areas in the vicinity of the Kigali International Airport (located approximately 5 km to the east of the *Nyarugenge* hill) (ibidem).

In 1990, Kigali already had a population of more than 200,000 and many governmental and administrative buildings have been added to the urban infrastructure (Goodfellow & Smith 2013). A decision to expand the city borders once again, to 112 km², and to establish the Kigali City Prefecture (*Préfecture de la Ville de Kigali*) in order to ensure efficient organisation of the capital's life and development, was also made in the same year (Manirakiza 2014; Benken 2017; Manirakiza et al. 2019). Despite its considerable population size and important administrative role of a capital city, in the early 1990s, Kigali largely consisted of massive, densely developed informal housing estates, populated by people migrating from the provinces. For this reason, on the eve of the genocide in 1994, the capital city of Rwanda was described as "really like a big village" (Goodfellow & Smith 2013, p. 3188).

5.1.4 The genocide and the period after 1994

Already at the beginning of its existence, the Kingdom of Rwanda was an example of one of many strictly hierarchical kingdoms in the Great African Lakes region. The social structure was based on the division into three groups – Bantu farmers traditionally inhabiting those lands called Hutu; Nilotic cattle breeders from the north, that is, Tutsi; and the least numerous, marginalised Twa group (Kuperman 2000; Tadjo 2010).[8] Throughout the centuries preceding the arrival of the Europeans, this social structure had developed in such a way that sometimes membership in a given social class was determined by assets in the form of a cattle herd. For example, a successful Hutu farmer (who has, for instance, earned enough to buy several animals) could become a member of Tutsi elites, while a Tutsi, having lost their wealth, became a Hutu (Tadjo 2010). Due to the initial misunderstanding of this complex structure on the part of the Belgian colonial administration, which applied the "divide and conquer" rule, the animosities between the two dominant groups aggravated over time under European occupation (ibidem). In an attempt to impose a superficial order that would suit their own needs, Belgians started issuing identity documents to every resident of the colony. Those documents specified whether a person belonged to the Hutu, Tutsi or Twa group on the basis of predetermined physiognomic traits (height, nose length and width, forehead height) instead of the economic status, which contributed to the later genocide. During their rule, Belgians favoured the members of the Tutsi minority; they appointed them to certain positions in the colonial administration while simultaneously stigmatising them in the eyes of the Hutu majority. Over time, Hutu came to perceive Tutsi as pro-European collaborators.

After decades of favouring Tutsi, at the end of the 1950s, before Rwanda gained independence, Belgians changed their policy to one which might have appeared more democratic and extended their support to the previously humiliated and discriminated Hutu, who made up 85% of the Rwandan society (Musahara & Huggins

2005; Goodfellow & Smith 2013). When Rwanda was established in 1962, the handover of power to the Hutu majority resulted in a policy of exclusion and stigmatisation of Tutsi. The group was also accused of causing more severe poverty in rural areas, which, in fact, was due to inefficient economic reforms. Over time, such measures led to the escalation of violence and power falling into the hands of an extreme organisation – the Hutu Power Movement, also called *akazu* (literally, a small house), which strived to completely exterminate Tutsi, whom they considered the chief enemies of the nation. Despite the presence of UN troops in Rwanda, on 6 April 1994, a plane with presidents of Rwanda and Burundi (both of whom were Hutu) was shot down. Although the assassination was carried out by a radical Hutu wing in order to rid the party of politicians who were willing to reconcile with Tutsi, it was the Tutsi group that was blamed for the attack. It was the direct cause of the first acts of genocide (Kuperman 2000; Tadjo 2010). Over the next 100 days, *Interahamwe* and *Impuzamugambi* (Hutu militants) carried out an organised extermination of approximately one million Tutsi, which was stopped by an intervention of Tutsi troops trained in nearby Uganda as part of the Rwandan Patriotic Front led by Paul Kagame. From the perspective of this book, it is crucial that researchers dealing with the Rwandan genocide also include those according to whom one of its causes – apart from the antagonisms between the two main groups inspired by Belgians – is also overpopulation and fights for the land (Uvin 1998; Lemarchand 2012). Being one of the smallest and most densely populated countries in Africa, Rwanda struggled with a shortage of spatial resources in the face of a dynamic population increase after the country gained independence. The colonial rule had led to numerous conflicts over land, quenched with violence during the time of the European domination. Amidst escalation of ethnic violence after the country gained independence, those conflicts were additionally aggravated by the inexperienced Hutu administration, which had in no way been prepared for power takeover. Land grabs became one of the motives behind killing members of the Tutsi group, in particular in rural areas (Lemarchand 2012).

The quantitative data on spatial planning in Kigali in the years 1994–1996 are extremely limited. During the genocide, the city became deserted, while public infrastructure and residential buildings were devastated. Right after the fights in Kigali ceased, the streets were empty and the houses stood abandoned (Goodfellow & Smith 2013). The city's population decreased significantly during the genocide. This reduction was caused, on the one hand, by the extermination of the majority of Tutsi inhabiting the capital (the Kigali Genocide Memorial alone, located in the city centre, is a tomb to approximately 250,000 residents of Kigali and nearby areas murdered during that time [Sodaro 2011]), and on the other hand – by some of Hutu fleeing to the neighbouring Democratic Republic of the Congo. At the end of 1994, the city's estimated population amounted to approximately 50,000 (Gazel et al. 2010, as quoted in Manirakiza 2014). Nevertheless, it should be noted that immediately after the genocide (already in July 1994), the completely deserted city experienced a population boom. Kigali was the safest and the most stable place in Rwanda, guarded by the troops of the Rwandan Patriotic Front. For this reason, it was quickly settled by Tutsi from all parts of the country who had survived the genocide, as well as refugees returning from abroad (mainly Uganda and the Democratic Republic of the Congo), who occupied the abandoned buildings already a few weeks after the fights had ceased (Goodfellow & Smith 2013; Manirakiza 2014; Benken 2017; Manirakiza et al. 2019). An additional contributor to the dynamic population growth in Kigali after the genocide was the 1996 decision of the Tutsi government to forcefully repatriate Hutu representatives hiding in refugee camps, mainly in Goma on the other side of the border with the Democratic Republic of the Congo. Approximately two million returnees were forcefully resettled by the army back to Rwanda. The majority of them decided to settle down in cities, including in particular Kigali, counting on more anonymity (Goodfellow & Smith 2013; Manirakiza 2014; Benken 2017; Manirakiza et al. 2019). The presence of the hostile Tutsi and Hutu groups, which were forced not only to live next to each other, but also to pursue the rights to the houses they occupied, led to numerous conflicts over spatial resources in Kigali (Musahara & Huggins 2005; Goodfellow & Smith 2013; Manirakiza 2014). Therefore, the city's authorities faced a number of challenges that were additionally amplified by a population boom typical of metropolises in Sub-Saharan Africa[9], as well as further intensified by migrations from the countryside to the city, which resulted in mass informal development of the suburbia (Benken 2017; Manirakiza et al. 2019). In the years 1996–1998, Kigali experienced chaos related to returning refugees, tense relations between the residents and high crime rates, which impeded a systemic approach to spatial planning (Goodfellow & Smith 2013). Moreover, food security of the ever-growing city population was also a serious problem. Due to restricted food supplies from rural areas, the residents were forced to farm land on parcels unoccupied by buildings, primarily those in marshy valleys and on steep slopes.

Already at the beginning of the next millennium, Kigali could hardly continue to be considered a "big village". Both its area and its population had increased threefold since the mid-1980s (Oz Architecture 2007). Due to the sprawl of buildings onto more and more hills, the city's authorities decided to take a step towards decentralisation. In 2000, the administrative borders of Rwanda's capital city were expanded to 314 km² and the Kigali City Municipal Authority (*Mairie de la Ville de Kigali*) was established in place of the Kigali City Prefecture. In 2002, the city's population amounted to approximately 603,000 with a population density of 1,924 people/km² (National Institute of Statistics of Rwanda 2002). However, investments in infrastructure development were unable to meet the needs of the growing population (Tsinda et al. 2013). In 2002, 70% of the city's area was occupied by informal buildings inhabited by the poorest residents (Manirakiza et al. 2019). In 2005, another administrative reform was carried out, expanding Kigali's borders to include vast urban-rural areas. Since that time, the total surface area of the Rwandan capital has amounted to 730 km². In the same year, the official name of the administrative division unit also changed from the Kigali City Municipal Authority to the City of Kigali (ibidem).

An important step towards restraining the uncontrolled development of Kigali and ensuring access to basic infrastructure for the residents (running water, sewage system, electricity, medical and educational facilities) was the implementation, in

2007, of a planning document entitled the Kigali Conceptual Master Plan, developed in cooperation with an American company, Oz Architecture. The objective of the new Master Plan was "to ensure sustainable urban development by using Rwanda's natural and human resources through the balancing of ecology, equity, and economy" (Manirakiza 2014, p. 169). The document assumed, among others, the establishment of a series of functional sub-centres that would take the pressure off of the traditional Central Business District (CBD) situated on the *Nyarugenge* hill. The first one is the Kimihurura Urban Center, a business and service hub with an area of 90 ha, located to the east of the *Nyarugenge* hill. It is currently characterised by a high share of urban greenery, exemplified by two large parks – the Kimihurura Park and the Kigali Centenary Park. Moreover, it also features the buildings of governmental administration, such as the Ministry of Justice, the Ministry of Foreign Affairs and International Cooperation as well as the Ministry of Defence, along with luxurious hotels (e.g. Radisson Blu Hotel & Convention Centre) and restaurants. Due to the functions fulfilled and its representative role, the *Kimihurura* hub has a small share of agricultural zones. Another sub-centre is the Kinyinya Town Center in the northern part of the city, covering an area of 200 ha. It fulfils service functions and is intended for the residents of the surrounding residential zone, Kinyinya Township. Apart from the sub-centres, the document also assumes establishment of the following residential zones: Niboye Sub Area, spanning 900 ha, situated to the south-west of the Kigali International Airport; Masaka Sector with an area of 4,500 ha in suburban zones in the south-east part of Kigali; and the *Rebero* leisure and conference centre with an area of 90 ha, located in suburban zones to the south of the city (Rubinoff 2011, as quoted in Manirakiza 2014).

The construction of the aforementioned service and business sub-centres required "clearing out" of vast areas covered by informal buildings, which involved large-scale resettlements of the poorest population. Due to the fact that Kigali lacks formal housing estates where displaced persons could be relocated, residents deprived of homes were forced to settle down far from the central city districts. In the suburbs, they once again occupied low-standard makeshift buildings, frequently located on steep inclines. Thus, the resettlements led to the issue of informal buildings being moved from the centre of Kigali to its outskirts, which additionally limited the poorest residents' options of finding gainful employment, further deteriorating their difficult situation. According to the assumptions of the Master Plan, new housing estates were constructed within the city. They were mainly intended for the elites from the emerging Rwandan middle class. Although the city's population density was successfully reduced thanks to those measures, the presence of new formal housing estates within the urban space additionally aggravated the issue of socioeconomic spatial segregation (Goodfellow & Smith 2013).

This is a good moment to discuss a new facet of comparisons between two of the analysed cases, Kigali and Singapore – namely, contemporary directions of spatial development pursued by the authorities of both cities. The Rwandan capital being called "the Singapore of Africa" is not unjustified. The next planning document for the capital city of Rwanda – Kigali Master Plan 2013 – was developed jointly with foreign experts associated with a Singaporean consulting company, Surbana

Jurong Private Limited, which provides services within the scope of infrastructure and planning in urban areas and develops planning documents for various cities around the world, including Singapore itself (City of Kigali 2013). The same company also participated in developing detailed local spatial plans after the previous Master Plan – Kigali Conceptual Master Plan – was published in 2007 by an American company called Oz Architecture (Goodfellow 2014). The Master Plan from 2013 introduced a division of the city into zones, regulating issues such as the manner of land use, the height of buildings and the permitted population density for each parcel. One of the document's features is also preserving the historical role of the CBD, developed on the *Nyarugenge* hill, as well as expansion of its range and increase of its multifunctionality. According to the 2013 Master Plan, the concise area of the city was divided into: (1) the historical business district of *Nyarugenge* with an area of 455.5 ha featuring the *Kivoyu* housing estate for the elites; (2) a new business district of *Muhima* with an area of 373.4 ha, featuring a zone dedicated to the development of advanced financial services and startups; as well as (3) *Kimicanga* with an area of 64.7 ha, which mainly fulfils recreational functions (City of Kigali 2013).

Another important assumption of the document is the organised resettlement of the population inhabiting areas with a steep terrain incline that are prone to landslides. Urban agriculture also amplifies dangerous mass wasting. Crop cultivation on steep slopes, without applying appropriate measures to stabilise them, intensifies erosion. The actions of the authorities aimed at limiting the use of zones with a steep incline should be considered justified as they protect the population from negative consequences of extreme mass wasting. However, one should note the serious negative socioeconomic consequences for the resettled population entailed by such top-down decisions of the authorities. First and foremost, they include loss of livelihood, reduction of food security, social bonds being severed, amplified social marginalisation and homelessness (Nikuze et al. 2018). The resettlements typically affect the most impoverished social groups, which are therefore the most vulnerable to marginalisation and cannot afford to live in better districts. The population is relocated to newly planned housing estates situated in the city outskirts. Although it means moving to housing estates with higher living standards compared with the previous informal, unsafe buildings on steep slopes, the new housing estates are frequently poorly connected with the remaining parts of the city. This creates a barrier to continuation of previous employment, which, in turn, considerably limits the options of securing a living. The situation is particularly difficult for people engaged in urban farming prior to the relocation. In the new place, the resettled individuals frequently lose access to land they could farm, which significantly limits their food security. In turn, finding a new source of income in a country experiencing a demographic boom and overpopulation is extremely difficult. What is more, establishment of new social relations between the resettled people is a complicated and time-consuming process.

One of the main concerns expressed by the critics of the 2013 Master Plan (e.g. Benken 2017) is the fact that none of the districts planned in the document was allocated to the construction of cheap housing estates for the poorest residents (ibidem).

T. Goodfellow (2017) noted that despite being aware of the problem, the authorities failed to take appropriate steps towards building affordable housing, which additionally exacerbates marginalisation of the poorest residents. Approximately 54% of Kigali's population is impoverished, that is, earning from USD 38 to 225 per month. Moreover, 13% of the city's residents live in extreme poverty, making less than USD 38 per month = approximately USD 1 per day (Uwayezu & Vries 2020). For that reason, the majority of Kigali's residents cannot buy or build a house using formal means and are somewhat forced to settle down in makeshift buildings in informal housing zones. Such buildings, located in areas not covered by planning documents, occupy up to 60% of space taken up by all housing areas in the city (ibidem). In 2012, nearly 63% of the city's population lived precisely in housing areas of this type (Tsinda et al. 2013). Informal housing affects in particular people who have a customary ownership right to the land or migrants from the countryside. Makeshift buildings are made of low-quality materials and typically have extremely low sanitary standards. After a countrywide construction law was introduced in 2007, use of traditional sun-dried mud bricks called *adobe* was prohibited along with the commonly used "wattle and daub" building method that involves using a wooden wall structure filled with or covered by mud and clay. The requirements contained in the construction law are at odds with reality, since the majority of current buildings in Kigali as well as those still being constructed are houses of this very type (Ilberg 2008). According to the Master Plan published in 2013, informal buildings, especially those located in floodplains and on steep hill slopes should be demolished. Due to the difficult situation in the housing sector in Kigali, the Rwandan government is pursuing a number of other measures aimed at ensuring that the poorest residents have safe housing, with running water and electricity, while simultaneously limiting the informal residential development. The authorities are implementing preferential conditions for developers investing in cheap apartments as well as low-interest housing loans for both investors and buyers. What is more, investors involved in the construction of cheap housing estates can count on support when it comes to land purchase as a result of expropriations. The government, on the other hand, undertakes to ensure availability of auxiliary infrastructure (roads, electricity, sewage system) (Uwayezu & Vries 2020). From the perspective of the year 2023, all of the aforementioned measures by the government have so far proven insufficient. Research carried out at a household level by E. Uwayezu and W. T. D. Vries (2020) proves that the majority of low-income residents, who are the target group of the government's projects, still cannot afford the new apartments sold on the formal market due to the high prices imposed by developers. One of the assumptions of sustainable development and the "right to the city" concept, represented by researchers such as H. Lefebvre (1968) and D. Harvey (2008), is high importance of inclusive spatial planning. Its elements include the measures taken by the government and city authorities aimed at ensuring that the poorest social groups are provided with housing that corresponds to both their needs and financial capabilities. In the case of Kigali, the poorest residents are therefore refused the aforementioned right to the city. On the one hand, they are being resettled from their previous homes, and on the other hand – they cannot afford to live in housing estates planned for them. This situation forces people to move to increasingly disadvantageous and remote locations, exacerbating their poverty.

Due to the aforementioned controversies around resettlements of people, the city's authorities have developed another updated version of the 2013 Master Plan. It takes into account adjustments from the seven-year period of implementation of the previous document as well as conclusions drawn from its evaluation. The basis for the new Master Plan, presenting a vision of Kigali's development for the years 2020–2050, is inclusion of the changing socioeconomic context as well as the predicted evolution of the residents' needs. The new plan introduces a number of modifications in relation to the previous document. First, it takes into account the possibility of individual buildings having different functions (residential, service, commercial), which increases the multifunctionality of space, while at the same time leaving the residents the freedom to integrate many forms of business activity. Moreover, small offices can be located in residential districts without the need to lease expensive office space in the CBD of *Nyarugenge* or the *Kimihurura* sub-centre. Second, the new Master Plan highlights two features of the spatial and functional structure which were considered typical of a sustainable city by P. Bächtold (2013) – compactness, expressed by building and population density, as well as proximity, understood as the distance between areas fulfilling different functions. These are supposed to be the future attributes of Rwanda's capital city. According to the document's assumptions, the population density of Kigali by 2050 is supposed to increase nearly threefold, from the current 1,778 to 5,243 people/ km^2 in the middle of the century. Analogically, the minimum number of residential buildings per hectare is supposed to be 70 instead of the current 25. Such decisions are aimed at more efficient use of space, taking into account the anticipated further increase in the number of residents. However, higher population density is supposed to be achieved not by building skyscrapers, as assumed in the previous document, but by constructing more mid-rise buildings in place of the informal single-storey ones. What is more, as many services as possible are supposed to be provided locally so that the population is not forced to travel far, intensifying the traffic on already congested streets of Kigali. In line with the new plan, every resident should have access to basic infrastructure and service facilities (e.g. banks, restaurants, stores, schools, offices and religious sites) within a short walking distance (Nkurunziza 2020). Measures supporting location of various forms of business activity within a single housing estate increase its functional heterogeneity and considerably decrease the distance covered by its residents within the city space. The 2020 Master Plan also envisions decentralisation of business activity within the city by developing and consolidating new service and commercial sub-centres in order to take off the pressure from the existing traditional business district on the *Nyarugenge* hill. These sub-centres are *Kinyinya* (to the north of *Nyarugenge*), *Kimironko* (to the west of *Nyarugenge*), *Kicukiro* (to the south-east of *Nyarugenge*) and *Nyamirambo* (to the south of *Nyarugenge*). Plans also include establishment of regional industrial centres away from the residential districts – *Ndera* and *Rusororo* (on the north-east outskirts of the city), *Masaka* (in the south-east) and *Gahanga* (in the south). All listed sub-centres and regional centres are supposed to

Figure 5.1 An estate of detached houses planned on a top-down basis in the *Kicukiro* district, with small maize fields visible between the buildings.

Source: Photograph by A. Górna, July 2019.

be connected by corridors as part of the planned Bus Rapid Transport network, modelled on Singapore's transport network (City of Kigali 2020).

Since its establishment, Kigali has been characterised by a concentric spatial development model, with the oldest, CBD located on the *Nyarugenge* hill. It was surrounded by the residential zone for lower and middle social classes, while river valleys were occupied by vast agricultural zones. In the outskirts, on the other hand, modern housing estates inhabited by upper classes were located (Manirakiza 2014). After Rwanda gained independence, in particular after the 1994 genocide, the city experienced intense population growth associated with uncontrolled, spontaneous expansion of informal residential buildings in marshy areas, as well as on steep slopes. In early 20th century, the city authorities adopted a policy aimed at organising the previously chaotic spatial development of the city. Subsequent published Master Plans assumed in particular decentralisation of the business and service activity and establishment of new sub-centres outside the *Nyarugenge* hill as well as construction of new housing estates, planned on a top-down basis (see Figure 5.1), whose presence was meant to limit the spread of marginal districts. However, the actions taken by the authorities proved insufficient and could not increase the inclusiveness of the residential sector (Manirakiza et al. 2019). According to the data published by the City of Kigali (2020), it is estimated that at the moment (data for 2020), more than 60% of developed areas in the city are still informal districts (see Figure 5.2) inhabited by 60% of the city's population. Informal buildings are frequently located at the bottom of the marshy valleys, which are

Figure 5.2 Informal buildings on a hill slope in the *Gasabo* district.

Source: Photograph by A. Górna, July 2019.

used for agricultural purposes. Their development leads to reduction of the area of agricultural land in the city.

However, the beginning of the 20th century was also a period when the development of Kigali was supported by the funds flowing in as part of foreign developmental aid (Goodfellow & Smith 2013). The authorities of Rwanda introduced a number of solutions aimed at attracting foreign investors to the city. Thanks to the governmental incentives and stabilised political situation, the capital city of Rwanda became a very advantageous location for investors (mainly from the tourist, financial, power and communications sectors), among others, from Canada, the United Arab Emirates or China. Modern buildings housing the headquarters of foreign companies and organisations (ibidem), such as Radisson Blu Hotel & Convention Centre, Kigali Marriott, Ecobank and the International Monetary Fund, began to appear in the city. As a result, the contemporary urban mosaic of Kigali is a combination of modern business districts (see Figure 5.3), luxurious hotels and apartments as well as organised upper-class housing estates, planned on a top-down basis, which contrast with the nearby continuously expanding makeshift houses of the poorest population (Manirakiza 2014; Manirakiza et al. 2019) (see Figure 5.4).

The capital city of Rwanda is characterised by heterogeneity typical of urban landscape, where areas with extremely varied building standards and wealth levels of the residents are adjacent to one another (see, for instance, Figure 5.10). This phenomenon is additionally amplified by the terrain. Flat hilltops are where districts for the middle and upper classes are being planned, while the slopes and the

Figure 5.3 Modern buildings in the central business district of *Nyarugenge*.
Source: Photograph by A. Górna, July 2019.

Figure 5.4 A typical single-storey informal building in the *Kicukiro* district.
Source: Photograph by A. Górna, July 2019.

bottom of the valleys are occupied by informal buildings of more impoverished social classes as well as vast farming zones. The socioeconomic disproportions, uncontrolled expansion of the city and neglected infrastructure continue to present extremely serious challenges for the authorities, which are striving to put Kigali on the track to sustainable development.

5.1.5 Secure urbanisation and green reforms

Despite the serious problems facing the dynamically developing capital city of Rwanda, modern-day Kigali undoubtedly stands out among other cities of the continent in terms of two aspects – secure urbanisation (Goodfellow & Smith 2013) and advancement of measures aimed at protection of the natural environment. One of the main premises of the 2020 Master Plan is the so-called green growth. By 2050, Kigali is supposed to become a metropolis designed with the natural environment and well-being of the citizens in mind. The city is supposed to develop based on rational use of the advantages of ecological systems while simultaneously protecting those systems for future generations (City of Kigali 2020).

The first thing of note is the distinct transformation from a city plagued by chaos and violence following the 1994 genocide to one of the safest metropolises on the continent. The city's authorities, instead of investing in an increased number of police officers, adopted an approach based on a high degree of social participation in ensuring safety in the streets (Barihuta 2017). It is the citizens (especially the youth) that patrol the streets, in exchange for which they receive small remuneration from other residents (Goodfellow & Smith 2013). The effectiveness of participatory policy is exemplified by the development of a form of public transport, the so-called mototaxi. In the 1990s and in the early 21st century, the mototaxi system was closely connected with organised crime, including drug trafficking. The government took firm steps towards solving the problem by introducing strict control over the mototaxi network and establishing an internal security department, which included mototaxi drivers themselves. They were incorporated into the urban street security system by being made directly responsible for providing protection to the residents. The mototaxi network is only one of many examples of how the residents' involvement in ensuring order in Kigali allowed for effective curbing of street crime (ibidem).

The government's policy in terms of ensuring clean streets and fighting used plastic is also important from the point of view of sustainable development. In 2008, Rwanda introduced a complete ban on production, sale, import and even use of plastic bags (Xanthos & Walker 2017). Thus, it went a step further than other countries on the continent, such as South Africa in 2003 and Tanzania in 2006, where use of plastic bags thinner than 30 μm was banned, as well as Uganda, where a fee of 5 cents per piece was imposed on plastic bags in 2007 (ibidem). Rwanda's case is an example of extremely radical anti-plastic policy. According to the law, on every last Saturday of the month, citizens as well as individuals staying in the country, including tourists, have a duty to participate in obligatory unpaid social work (Hasselskog & Schierenbeck 2015). The initiative is described as *Umuganda*, which translates from the Kinyarwanda languages as "a meeting with a common goal" or "we are working together" (Barnhart 2011, p. 8). However, this word, which dates back to the pre-colonial times, has negative connotations, since during the colonial era as well as the early years of independence, it was used to describe forced labour aimed at strengthening control over Rwandans (Uwimbabazi & Lawrence 2013). Currently, *Umuganda*, as a means of implementing the government's policy to eliminate poverty, promote unity and reconciliation, build interpersonal bonds as well as introduce decentralisation, includes a number of works consisting in building and maintaining roads, planting trees or communal cleaning and patrolling of streets (Goodfellow & Smith 2013; Uwimbabazi & Lawrence 2013; Hasselskog & Schierenbeck 2015). Although the last activity has been criticised multiple times due to being compulsory, it may be the reason why the capital city of Rwanda is nowadays called "the cleanest city in Africa" (Twahirwa 2018).

In 2006, the authorities of Kigali also commenced measures aimed at protecting the environmentally valuable urban wetlands at the bottom of the valleys. According to the 2013 Kigali Master Plan, construction within a radius of 20 m from wetlands in designated ecologically sensitive zones was prohibited (City of Kigali 2013). All buildings which are currently located in close proximity to the marshlands are supposed to be removed and their residents are to be relocated. According to the master plans from 2013 and 2020, wetlands are meant to be used for tourism and recreation purposes. There are plans to establish, among others, two public parks: Kigali CBD Wetland Park with an area of nearly 178 ha, located in a valley on the border of three districts – *Nyarugenge, Gasabo* and *Kicukiro*, as well as the Nyandungu Wetland Park with an area of 130 ha, located near the airport. Moreover, the largest wetland recultivation project in the capital of Rwanda is supposed to be carried out on the southern edges of Kigali, near Lake Muhazi. In the future, these ecologically valuable areas are meant to become the main tourist and recreational destination outside the city (City of Kigali 2020).

One of the city's serious problems is air pollution, mainly associated with transport (Nduwayezu et al. 2015; Nahayo et al. 2019). The city is dominated by poorly maintained old cars, motorbikes and scooters, which produce excessive amounts of harmful exhaust fumes. Other sources of pollution include open hearths and low-quality furnaces, wood- or charcoal-fired, which are commonly used especially by the poorest residents (Henninger 2009), as well as heavy industry (City of Kigali 2020). The city also lacks a proper monitoring system that would allow for controlling the level of harmful substances (Nahayo et al. 2019). Emission of pollutants, especially carbon dioxide, carbon monoxide and particulate matter (PM10 and PM2.5), considerably exceeds the norms recommended by the World Health Organisation, putting the health of the residents at risk (Henninger 2009; Henninger 2013; Nahayo et al. 2019). Varied terrain results in smog accumulating primarily at the bottom of the valleys (Henninger 2009; Henninger 2013; Subramanian et al. 2020) where agricultural activity is pursued. In addition, those valleys, which are the least suitable locations for development, are inhabited by the poorest residents of the Rwandan capital. As a result, those social groups are also the most vulnerable to negative effects of excessive exposure to harmful substances contained in the air, and over time also accumulating in the soil used for growing plants. Air pollution directly affects the quality of the food produced since it is absorbed by the plant leaves. Vegetables grown at the bottom of the valleys are sold at local markets and in supermarkets or end up in hotels and restaurants, thus posing a risk to the health of consumers from all Kigali social groups, not only those inhabiting the valleys. According to the 2020 Master Plan, all heavy industry facilities are supposed to be relocated to the city outskirts, away from residential areas (City of Kigali 2020). The document also introduces a number of measures aimed at limiting smog from transport. They include, among others: introduction of strict control over import of vehicles, investments in the public transport system, as well as promotion of planning of multifunctional housing estates, limiting long-distance transport of people and development of cycle lanes (ibidem). Other steps towards improving air quality planned by the authorities of Kigali include protection of natural marshlands, increasing the share of urban greenery within the city space, as well as planting trees (City of Kigali 2020). According to J. B. Nduwayezu et al. (2015), tree planting and introducing a pollution tax can be effective measures in fighting fume emissions. The 2020 Master Plan envisages tax reliefs and financial support for all energy-saving initiatives (City of Kigali 2020). Moreover, the document also includes increased use of renewable energy sources (up to 20%); however, it does not contain any more detailed information on how the assumed level is to be achieved. For that reason, the provisions of the Master Plan give rise to certain doubts as to actual abandonment of fossil fuels and charcoal. The research conducted by J. L. Seburanga et al. (2014) creates slightly more hope as regards the effectiveness of measures pursued by the authorities. It demonstrates that at housing estates planned on a top-down basis, the share of area covered by trees is as much as 35%, more than four times higher than in informal districts, which are characterised by much higher building density and for which the share of green areas does not exceed 8.5%. Therefore, even though replacement of informal housing estates with new ones is controversial due to resettlements of the population, it may be justified in the context of improving the quality of the natural environment. The direction of spatial planning chosen by the Kigali authorities, aligned with the concept of sustainable development, may therefore entail benefits for the urban ecosystem and the quality of life of the residents, assuming that the provisions of the planning documents will be effectively implemented and successively evaluated.

5.2 Institutional and legal framework and the authorities' policy with respect to urban agriculture

The location and the terrain clearly determine the distribution of urban agriculture in Kigali. On the other hand, the landscape had limited impact in the case of other cities analysed in this book. Other factors important for the spatial distribution and development of the role of agriculture within the spatial and functional structure – similar to Havana and Singapore – include the institutional and legal framework as well as the attitude of Rwandan authorities to agricultural activity within the city borders. They are reflected, among others, in the strategic documents described above.

The effectiveness of spatial planning is directly affected by the land ownership system in Rwanda. During the period immediately following the genocide, in the face of returning refugees settling down back in Kigali, the city's authorities struggled with the issue of claims to land intended for development. Due to its insufficient supply and the land distribution system in force at the time, the needs of the dynamically growing city population could not be met. The people arriving in the city were forced to occupy lands owned by the state without having obtained the usufruct or property rights (Bizimana et al. 2012). This situation persisted for a decade, until 2005, when Organic Law was introduced (Government of Rwanda 2005). It guaranteed an opportunity to purchase land and obtain the usufruct right regardless of ethnicity or gender (Government of Rwanda 2005, Art. 4). At the time, the law protected land acquired in the customary manner and land purchased using legal means to an equal extent. According to E. Uwayezu and W. T. D. Vries (2020), as a result of the reform, the outcome of which was the Organic Law, a specific system of statutory ownership right developed in Rwanda. It distinguishes two main categories of land: state-owned and private. State-owned land includes wastelands, forests, wetlands, parks, reserves and land occupied by public institutions. Private land, on the other hand, belongs to citizens. In reality, however, natural persons own land not based on ownership right, but based on a renewable, long-term (emphyteutic) lease agreement. Organic Law enabled the authorities to carry out expropriation due to public interest upon prior adequate compensation (Government of Rwanda 2005, Art. 3). This provision, assigning extremely broad competences to state officials, was highly controversial, in particular due to a flexible approach to which projects were considered public interest projects (Goodfellow 2014). Another law, determining the conditions of managing land resources, known as the Land Governing Law, established the necessity to register land purchase according to the Land Tenure Regularization (LTR) programme (Government of Rwanda 2013). Although the state authorities are not the main land owner, in Rwanda, similar to Havana or Singapore, the entire process of its management is subject to strict control of national institutions, which are simultaneously the main decision-makers in this regard.

An important factor affecting the presence of urban agriculture in the space of Kigali was the threat to the city residents' food security following the genocide. In the early 1990s, the agricultural sector in Rwanda was underdeveloped and the obsolete agricultural tools and techniques considerably limited its productivity. Moreover, intensified migrations to the city contributed to overpopulation. As a result, the most impoverished people were forced to build their houses and settle down in areas occupied by agriculture. This reduced the area of agricultural land and, as a consequence, seriously limited agricultural production within the city. Additionally, due to the unstable situation in the country and resettlements of the farmers, agriculture in rural areas was also facing a serious crisis. It resulted in additional reduction of food supplies to cities. In the early 21st century, Rwanda's Ministry of Agriculture, supported by foreign donors and the Food and Agriculture

Organisation of the United Nations (FAO), took steps to achieve food security of the residents (Kinka et al. 2014). Implementation of the Strategic Plan of the Transformation of Agriculture Phase I (PSTA), which was supposed to considerably improve the situation in Kigali, commenced in 2004. Modernisation of the infrastructure associated with food storage allowed for reduction of seasonal food shortages. The plan also assumed increased monitoring of the residents' food security and promotion of a healthy and varied diet among city inhabitants. Among others, the Food Security Information System was established, which helped in collection of data on food availability and food security in different regions of the country. Since the 1990s, most food was supplied to Kigali from rural areas; upon the FAO's recommendation, a decision was made to include urban agriculture in the city development strategy. It was included in the Kigali Master Plan published in 2009 (Kinka et al. 2014). Promotion of urban agriculture contributed to reduction of the distance between the place of production and the place of distribution and consumption of food, thanks to which its availability in the city improved. As a result, already in 2015, merely 3% of Kigali residents lacked food security, while the share of undernourished people in the city stood at 25%, making these values the lowest in all of Rwanda (MINAGRI 2016). However, it should be noted that according to the results of the research published by the Ministry of Agriculture, as much as 34.5% of the city's population still remains on the verge of food security (ibidem). The current situation in Kigali therefore remains unstable and additionally threatened by climate change, which can contribute to reduced agricultural yield and, as a consequence, limited food availability in the city.

In view of the successful solutions implemented in Rwanda's capital city, the authorities have somewhat changed their approach to urban agriculture. The document that will be shaping the distribution of agricultural activity within Kigali's space in the nearest future is the 2020 Master Plan. According to its assumptions, by 2050, the land occupied by agriculture is supposed to amount to 165 km^2 and make up 22.7% of the total area of the city. These values are lower than those stated in the previous version of the document from 2013, where they amounted to 192.9 km^2 and 26.4%, respectively (City of Kigali 2020). The priority measure in terms of agriculture is its modernisation and mechanisation, as well as increasing its productivity while simultaneously reducing the use of artificial fertilisers and pesticides. Due to inappropriate agrotechnology practices, arable land, primarily that located in the northern part of the *Gasabo* district, is severely degraded. One of the objectives of the Master Plan is recultivation and protection of agricultural areas, followed by promotion of sustainable production methods that would allow for preservation of high soil quality. Drawing up the Urban Agriculture Development Plan, Urban Agriculture Extension Manual and Integrated Urban Agriculture Management Plan is supposed to make implementation of the above assumptions possible. What is more, the document also sets out to organise a number of trainings to raise awareness of sustainable agricultural techniques addressed to the city residents. However, the provisions of the Master Plan are ambiguous. On the one hand, the document assumes integrated measures aimed at protecting urban agriculture in some of the marshy valleys and low-incline areas. It is supposed to be a part of the strategy for improving the residents' food security. Agricultural activity within the wetlands is also meant to purify greywater (dirty water that is free from faeces and can be reused after appropriate purification) from residential districts. However, crops in the majority of valleys within the city limits are supposed to be moved to highland areas to protect the environmentally valuable wetlands. The document also proclaims development of innovative, sustainable agricultural techniques that are optimal for land with an incline of more than 20%, as well as additional financial support for residents applying slope terracing. Crop cultivation in valleys is supposed to be replaced by small household gardens and rooftop farming. The Master Plan even assumes establishing so-called Zero Net Loss of Agricultural Areas zones within the city and integration of urban horticulture in urban design. Thus, the measures included in the document indicate the intention to preserve agriculture within Kigali's space. However, it turns out that the majority of the aforementioned provisions refer to rural areas within the administrative borders of Kigali. According to the city's division into zones corresponding to different forms of land use, agriculture is supposed to be preserved only in rural areas in the northern part of the *Gasabo* district and in the south of *Nyarugenge*. However, the planned share of agriculture in central parts of the city is minor. It is only meant to occupy some of the valleys in the *Kicukiro* district (to the south and north of the airport), on the border of densely developed areas (City of Kigali 2020). Taking into account the fact that crops are currently also present within the urbanised zone, even in central districts of the city, large-area agriculture in particular should be expected to be removed in the near future into more and more peripheral areas.

The authorities of Kigali have adopted a slightly different policy with respect to small crop fields. A common practice (almost identical to that pursued in Havana), mentioned in an interview by the representative of the Kigali City Hall, is granting a temporary right to lease unused parcels with a small area for food production purposes. Every resident that wants to use a field for farming (whether that field is state-owned or private) can report to the relevant department of the City Hall. Lease can be granted under two conditions. First, the selected parcel must be a wasteland. Second, agricultural activity pursued on it cannot be conducted on an industrial scale. Only subsistence farming is permitted, with small surpluses being sold at local markets. Of course, when the original owner of the parcel expresses the wish to develop it, the urban farmers must change the location. Such a practice allows for temporary utilisation of unused space while simultaneously supporting the local population by supplying it with food.[10]

Another practice aligned with the policy of combating malnutrition among the residents is promotion of the so-called kitchen gardens (Rw. *akarima k'igikoni*). The city authorities promote fruit and vegetable cultivation to meet the needs of individual households. A campaign entitled "Promoting Diversified Diet and Innovative Urban Farming for a better and Well-Nourished City", carried out in 2017 by the Kigali City Hall in cooperation with the International Potato Centre, is an example of campaigns pursued by the city authorities in order to improve food security of the residents as well as enhance and diversify their diet, especially among children. The campaign involved a number of trainings aimed at educating the population on organic fruit and vegetable cultivation along with healthy nutrition (Rwanda Water Portal 2021).

Based on the premises described in the implemented planning documents and based on the directions of development set out by Rwanda's government, changes in terms of distribution and characteristics of urban agriculture in Kigali should be anticipated. Taking into account the fact that Singaporean experts played an important role in developing the current Master Plan as well as its previous versions, it seems crucial to ask to what extent the measures regarding local food production will resemble those taken in the Asian city-state. Despite the fact that the city authorities are the main decision-makers shaping the spatial policy of the Rwandan capital, including spatial distribution of agricultural areas within its limits, the existing institutional and legal framework grants the residents a certain degree of agency in that regard. Thanks to the possibility of using wasteland zones for growing food crops, the population of Kigali has a chance to pursue a grassroots transformation of its immediate vicinity according to its needs, even if these changes are merely short term.

5.3   Distribution of agriculture within the urban space

The number of all agricultural areas identified in Kigali using Google Earth Pro tools amounted to 780. Ninety-eight of them were analysed in the field and it is those areas that will be subject to a detailed analysis in this chapter.[11] The largest number of agricultural zones – 48 – is located in the *Gasabo* district. Other districts feature less than a half of that number – 23 in *Kicukiro* and 17 in *Nyarugenge*. Moreover, five agricultural zones are located on the border between *Nyarugenge* and *Gasabo* districts, two between *Nyarugenge* and *Kicukiro* districts, while three are situated between *Gasabo* and *Kicukiro* districts. The following sub-chapter will present the differences in the area occupied by urban agriculture in individual districts.

The characteristics of distribution of urban agriculture in Kigali differ considerably from those in Havana or Singapore. Agricultural areas are present in all city districts; however, the factor that directly determines their location is the terrain, which was insignificant in the case of the other analysed cities, characterised by low terrain variability. Kigali is located on several densely developed hills. The height differences can amount to as much as 400 m. Urban agriculture wraps around each of the hills and unifies the urban tissue by using land that is difficult to develop due to disadvantageous water conditions or excessive incline. Taking into account the location within the city space and the area covered, the analysed agricultural zones can be divided into two major groups. The first one comprises large-area agricultural land occupying vast, marshy bottoms of the valleys that stretch between the hills. Out of the 98 analysed agricultural zones, 35 are located in the valleys and on their slopes. They cover vast areas and have an elongated, ribbon-like layout. During the rainy season, which in Rwanda lasts from October to May, the surface water level in the valleys visibly rises. Due to excessive moisture, use of that land for construction is therefore highly impeded. For that reason, it is

Figure 5.5 The bottom of a valley divided into smaller fields in the *Gasabo* district; two
greenhouses are visible at the bottom of the valley.

Source: Photograph by A. Górna, July 2019.

allocated to agricultural activity. Vast areas occupied by agriculture are frequently adjacent to densely built-over residential zones, occupying the slopes of the valleys and creating a clear-cut boundary in space (Figure 5.5).

Moreover, the vicinity of residential buildings and the bottom of the valleys used for agricultural purposes creates a distinct landscape of rural features converging with urban ones. This is typical even of the most central parts of Kigali. The agricultural areas are adjacent to both densely packed informal buildings, frequent in districts inhabited by the poorest residents, as well as slightly sparser detached houses owned by representatives of the middle and upper-middle classes. The densest network of valleys, that is, the largest number of valleys per area unit, is present in the *Gasabo* district, with agricultural areas surrounding *Kacyiru, Remera, Nyarutarama* and *Kimihurura* hills. *Nyarugenge* and *Kicukiro* feature a much sparser valley network, and therefore a lower share of agricultural zones within their space.

The second group of analysed agricultural areas comprises 50 parcels located on hill slopes and 13 situated on flat hilltops. They typically take the form of small arable fields or vegetable gardens between residential buildings and are present in all analysed districts. Figure 5.6 presents the distribution of urban agriculture within the space of Kigali, both the 98 analysed agricultural zones and the remaining 682, previously identified in Google Earth.

Figure 5.6 Land development and distribution of urban agriculture in Kigali.

Source: Own elaboration based on Górna and Górny 2021 (Creative Commons CC BY 4.0 license).

Figure 5.7 A crop field next to a house located on the edges of the research area.

Source: Photograph by A. Górna, July 2019.

The central parts of Kigali feature a much lower number of small arable fields on slopes and tops of elevations. Agriculture in those parts is mainly contained to the vast bottom of the valleys. This is primarily due to high building density in the central sectors. Small distances between the buildings limit the space that could be allocated to fields and vegetable gardens. The share of agricultural zones located on slopes and hilltops increases along with the distance from the centre of Kigali. It results from reduced building density towards the city outskirts. The largest number of small arable fields is located on the edges of the research area, characterised by the least compact spatial form (see Figure 5.7).

The spatial distribution of urban agriculture in Kigali reflects the nutritional needs of the residents. Therefore, it typically occupies areas in the immediate vicinity of residential buildings. On hill slopes and the flat terrain on the hilltops, they take the form of the aforementioned kitchen gardens. At the bottom of the valleys, crop fields are also adjacent to zones characterised by relative density of residential buildings (see Figure 5.5). It is less frequent for agriculture to be situated in places dominated by service and commercial or administrative functions in economically advantageous locations. The *Nyarugenge* and *Kimihurura* hills can serve as examples in that regard – they are developed to be prestigious service and business centres and residential zones for the elites. The number of crop fields is small in those areas and agriculture is mainly allocated to the bottom of the surrounding valleys. In those sectors, the share of agriculture in the yards of individual detached houses is also low despite it being a common practice in other districts.

Figure 5.8 A small wasteland zone allocated to yam and banana farming in the *Gitega* sector in the *Nyarugenge* district (right).

Source: Photograph by A. Górna, July 2019.

The key practice determining the ubiquity of urban agriculture in Kigali is the
possibility of allocating all wasteland zones to growing crops. Thanks to it, even
the smallest wasteland zones transformed into crop fields can be found within the
city space, as presented in Figure 5.8. Although the practice is based on a grassroots initiative and participation of the residents in shaping the city space, it also
entails some significant risk. That risk is agricultural use of areas that are unsuitable
for growing crops, for example, excessively tilted slopes. This leads to intensified
soil erosion and, as a consequence, also amplification of mass wasting (landslides,
downhill creep, soil being washed away) that is dangerous to the residents. Moreover, environmentally valuable wetlands at the bottom of the valleys are put at risk.
The foregoing problems were included in the directions of spatial planning set out
by the authorities of Kigali. According to the published Master Plans from 2009,
2013 and 2020, agriculture, especially large-area agriculture, is to be successively
removed from the city's central districts. The majority of the marshy bottoms of the
valleys, which are currently occupied by crops, are also supposed to be protected
as environmentally valuable areas, where agricultural activity is not envisaged. The
only exceptions are some of the agricultural areas located in valleys at the border
of the *Niboye* and *Kanombe* sectors and in the *Nyarugunga* sector in the *Kacukiro*
district. What is more, according to the current document, agricultural activity is
supposed to be limited on land characterised by a steep incline (exceeding 15%) in
order to prevent slope erosion.

A factor that may significantly contribute to intensified removal of agriculture from central districts of the city in the upcoming years is dynamic building development. Kigali is an example of a city characterised by a relatively high share of wasteland zones within its space. However, due to the increasingly stronger economic and political position of Rwanda as well as increased number of foreign investments, building density is being more and more frequently increased in sectors such as *Nyarugenge, Kimihurura, Kacyiru, Kinyinya* and *Remera*, while gaps in the urban tissue are being filled. Since those gaps have so far been occupied by crops and the rights to land use granted to residents are short term, the share of agricultural areas in the fastest-developing parts of Kigali is going to decrease (see Figure 5.9). One of the analysed agricultural areas located in the *Kinyinya* sector in the *Gasabo* district (see Figure 5.5) is an interesting example illustrating both the situation of urban agriculture and the changes taking place in Rwandan society. The satellite image from June 2019, available in Google Earth Pro, clearly shows crop fields covering an area of 23 ha. The presence of agriculture in this location was also confirmed during the field research in July 2019. However, it turns out that already in March 2020, the parcel had been nearly completely cleared out. It is currently occupied by a golf course. The spatial coverage of crop fields has been radically reduced. They have only been preserved on the edges of the golf course. Agriculture in this location currently covers an area of 4.9 ha. The fact that the crop fields have been replaced by a luxurious recreational facility demonstrates

Figure 5.9 Cassava, maize and sweet potato cultivation in a field located on a slope in the *Muhima* sector, *Nyarugenge* district. A parcel adjacent to the field is allocated to hotel construction.

Source: Photograph by A. Górna, July 2019.

the demand for similar forms of leisure and can be a sign of the Rwandan society becoming wealthier, as well as tourists and entrepreneurs coming from abroad. Removal of urban agriculture to city outskirts has been observed by M. Taguchi and G. Santini (2019). The authors emphasise that since the 1990s, when agriculture was included in Kigali's spatial planning policy pursuant to the advice received from the FAO, the city has undergone an extremely dynamic development. The Rwandan capital has become one of the fastest-developing cities in the world, while agriculture, slowly losing its importance, has begun to be replaced primarily by residential buildings.[12]

The location of urban agriculture has a significant impact on the production methods used and plant species grown, which will be discussed more broadly in the following sub-chapter.

5.4   Characteristics and functions of urban agriculture

5.4.1   *Structural and production characteristics*

Similar to Havana and Singapore, the area of all agricultural zones visited during the field research was estimated using the tools of Google Earth Pro. It amounts to 1,170.4 ha, that is, 12.7% of the research area, which covers 9,180 ha. The average surface area of a single agricultural zone is 11.9 ha. These values are considerably higher than in Havana, where the total area of gardens amounted to 32.3 ha (with an average of 0.75 ha), as well as in Singapore, where the total area stood at 118.6 ha (with an average of 5.24 ha).

The largest section of land allocated to agriculture in Kigali, situated in the southern part of the *Kicukiro* district, has an area of 284.6 ha (more than twice the total area of urban gardens and farms analysed in Havana and Singapore). Interestingly, it is simultaneously a zone where the agricultural function is supposed to be preserved according to the 2020 Master Plan. In turn, the area of the smallest household field, located in the *Nyarugenge* district, does not exceed 90 m². Taking into account both the agricultural areas analysed in the field and polygons identified exclusively on the orthophotomap, the total area of urban agriculture amounted to 1,382.1 ha, which accounts for 15.1% of the research zone.

In view of the differences in spatial distribution of agricultural zones in Kigali, the area occupied by urban agriculture in individual districts should be discussed. In *Gasabo*, the analysed agricultural zones take up 657.8 ha (15.2% of the research area within the district), with an average size of a single field amounting to 13.2 ha. In *Kicukiro*, agricultural zones have a larger average area of 15.6 ha; however, their total area is smaller, amounting to 435.5 ha (13.6% of the research area in the district). In turn, in *Nyarugenge*, agriculture only occupies 77.1 ha (4.6% of the research area within the district). What is more, the average size of the fields is more than four times smaller than in *Gasabo* and nearly five times smaller than in *Kicukiro*, as it only amounts to 3.2 ha. An analysis of both the area of agricultural zones examined in the field and those identified exclusively on the orthophotomap yields the following results: in *Gasabo*, the total area taken up by agriculture amounts to 743.8 ha

(17.1% of the research area within the district), in *Kicukiro* – 536.9 ha (16.8%) and in *Nyarugenge* – 101.4 ha (6.1%). The average surface area of all crop fields (analysed in the field and found exclusively on the orthophotomap) in *Gasabo* is 3.68 ha, in *Kicukiro* – 1.34 ha and in *Nyarugenge* – 1.26 ha. Such a huge difference between the average area of the fields in the entire research sample (780 agricultural zones) and the average area of those analysed in the field (98) results primarily from the fact that during the field research, access to small crop fields located between residential buildings was limited, so many similar areas were not included in the field research sample.

Nyarugenge has the smallest share of agricultural zones in its total area, primarily due to the fact that the district is dominated by administrative as well as service and commercial buildings, while its housing estates are intended mainly for residents with a higher social status. Given this functional layout, the share of household agriculture is small. Large-area agriculture is concentrated exclusively at the bottom of the valleys, which are situated on the border with other districts. Their coverage is limited in the *Nyarugenge* sector, which includes the CBD. The most common agricultural zones in the district are small crop fields between the buildings, located in its southern part, predominantly in the *Nyamirambo* sector, which is characterised by lower building density compared with other sectors of that district. In *Kicukiro* and *Gasabo*, the average area of agricultural zones is much bigger since those districts feature many more vast valleys allocated to crop cultivation. However, it should be noted that there are also more small crop fields in both districts than in *Nyarugenge*.

The analysed fields are clearly dominated by plant production. Animal breeding accompanied crop cultivation only in two instances – in one of those cases, the animals were turkeys, while in the other, they were goats. The most common plant species grown in Kigali are maize (identified in 68% of analysed agricultural zones), bananas (48%), cassava (39%), yams (37%) and sweet potatoes (24%). These food plants are typical agricultural plants grown throughout Rwanda; they form an important part of its residents' diet and play a key role in ensuring food security (Wambugu & Muthamia 2009). The remaining food plants identified during the field research included cabbage (11% of agricultural zones), potatoes (9%), rice (5%), tomatoes (3%), lettuce (3%), papaya (3%), eggplant (2%), amaranth (2%), onion (1%) and beans (1%). Twelve per cent of the analysed zones also featured fodder grass – a crop that is indirectly a food plant intended for fodder for farm animals.[13] Moreover, in four instances, decorative plants were also identified. In three of those cases, they accompanied cultivation of food plants. Decorative plant cultivation occupied the edges of the valleys.

A prominent feature of the structure of crops in Kigali is the small share of leafy greens (e.g. lettuce or spinach). Due to the fact that they spoil quickly, their cultivation is typical of urban areas, as observed in Havana and Singapore. Kigali, on the other hand, is dominated by species of grains (maize and rice), tubers (e.g. cassava, yams, sweet potatoes), as well as fruits (bananas, papaya). These plants are the core of people's diet, especially among the poorest Kigali residents. The city authorities have recognised the insufficient share of vegetables, especially leafy greens, as a major issue symptomatic of inadequate nutrition among the residents. One method of promoting a more diversified diet is the aforementioned campaign supporting establishment of kitchen gardens and cultivation of fruits and vegetables. When analysing the distribution of individual species of plants grown within the research area, it should be noted that due to the varied terrain, and therefore also varied soil and water conditions in Kigali, the location of agriculture has a significant impact on the production structure. The valleys are dominated by bananas, planted mainly on the slopes and on the edges of crop fields. They were identified in 83% of the analysed agricultural zones situated in the valleys. Other important crops grown in those areas include maize (68%), sweet potatoes (51%), yams (43%) and cassava (17%). Vegetables such as cabbage (34%), lettuce and tomatoes (9% each) and eggplant (4%) are grown much more frequently at the bottom of the valleys than on the slopes. What is more, the bottom of the valleys was where all rice crops identified during the field research were located. Only those marshy areas ensure water conditions that allow for the growth of rice, which is a hydrophilic plant. A distinct distribution of individual plant species was observed in the analysed valleys. Vegetables such as cabbage, lettuce or tomatoes and tubers – sweet potatoes, yams or cassava – were grown in central parts of the valley bottom. Maize crops were clustered closer to the slopes, while bananas were typically planted on the edges of the valleys and on the slopes (Figure 5.10). Furthermore, individual maize and yam

Figure 5.10 Maize and banana crops in a valley in the *Kimihurura* sector. The photograph shows distinct, diversified land use in a vertical layout: crops can be seen at the bottom of the valley, the closest buildings are informal ones, located on the slopes, while the hilltop is occupied by multi-storey service and business buildings.

Source: Photograph by A. Górna, July 2019.

plants were dispersed throughout the entire valley between other plant species. Such a spatial distribution recurred in the majority of the analysed valleys, as well as on the slopes, as presented on the valley cross-sections (see Figures 5.15–5.18).

The structure of crops grown on slopes and flat hilltops differs from that at the valley bottom. Apart from maize, whose share in this group of agricultural zones amounts to 68%, cassava was identified on more than 50% of fields in this group, while yams were on 33%. Cassava and yams are tuber species, resistant to water stress resulting from low level of groundwater, typical of Kigali hill slopes. Other tubers – sweet potatoes (9%) and potatoes (8%) – were also grown on the 63 fields situated on the hill slopes and tops; however, vegetables present in the valleys, such as cabbage, tomatoes or eggplants, have not been identified. Pumpkin and amaranth were grown on two fields. In three instances, tuber cultivation was accompanied by papayas and in two instances by mango trees. What is more, bananas have been identified on the slopes planted in free spaces between the buildings. However, their share was much lower than in the case of the bottom of the valleys and amounted to 29%.

It should be emphasised that nearly 79% of the analysed agricultural zones, both those located within valleys and those on slopes and tops of elevations, are used to grow more than one plant species, while 44% of them – more than two species.

The municipal authorities are also attempting to influence the structure of production and distribution of individual species of agricultural plants within the city. According to the 2013 Master Plan, so-called priority crops have been specified for two districts. For *Gasabo*, these were maize, vegetables, legumes, fruits, rice and soy, while for *Kicukiro* – coffee and fruits. According to the aforementioned document, agricultural activity in *Nyarugenge* was supposed to be restricted (City of Kigali 2013). Taking into account the fact that six years have passed between the publication of that version of the master plan and the conducted field research, certain effects of its implementation could already be expected. The document also assumed differences in the structure of the crops in individual districts. However, in reality, this structure is similar. Although *Gasabo* is indeed dominated by maize, distinguished in the document, the crop is also ubiquitous in other parts of the city. In *Kicukiro*, where coffee and fruits are supposed to be grown according to the Master Plan, maize, bananas and cassava also dominate. The situation is similar in *Nyarugenge* as well. However, it should be emphasised that the structure of the crops, in spite of being uniform in all districts within densely developed areas, may be more varied in the peripheries of the city (not covered by this analysis), primarily rural areas within the administrative borders of Kigali. According to the 2013 document and its updated 2020 version, agroforestry, which combines cultivation of trees and shrubs (including fruit-bearing ones) that stabilise the slopes with other plant species distributed between them (e.g. maize), is promoted in order to protect areas with an incline of 15%–25% against erosion. This practice is applied in some agricultural areas located on steep slopes of the elevations.

The main factor affecting the structure of crops is the residents' nutritional needs. Due to problems with ensuring food security in Kigali, the research area is dominated by basic food plants, such as maize, bananas or tubers. Perhaps the steps taken by the authorities to promote a more varied diet and increase fruit and vegetable production, especially in household gardens, will contribute to future changes in the structure of plant production in the city.

5.4.2 Organisational and technical characteristics

Apart from the structure of the crops, the location of agriculture also affects the applied production methods. Although plants are cultivated directly in the soil in all analysed agricultural zones, due to the characteristics of individual fields, it is necessary to apply appropriate agrotechnical measures. This concerns primarily the vast, marshy valley bottom. The high ground moisture content and clear increase of groundwater levels during the rainy season necessitate the use of raised plant beds for growing plants (see Figure 5.11), in some cases stabilised with wood fragments to prevent erosion. This does not concern maize and bananas, which are typically located in more external and less marshy parts of the valley bottom, as well as rice, which is grown on small fields flooded with water.

Due to the shallow presence of surface waters, small irrigation-drainage ditches that are permanently filled with water have been placed between the raised plant beds or individual fields. During the dry season, this water is used to irrigate the crops. Small buckets are typically used for this purpose to take the water directly from the ditches. On the other hand, during the wet season, the ducts drain excess rainwater. Within one of the analysed valleys, two greenhouses have been built

Figure 5.11 Vegetable and tuber cultivation on raised plant beds in the *Gasabo* district.

Source: Photograph by A. Górna, July 2019.

(see Figure 5.5). However, such structures have not been observed in other parts of the city.

In the case of agricultural zones located on slopes, in particular those with a steep incline, a commonly applied method that contributes to erosion reduction is terracing. However, use of this technique was not observed during the field research in Kigali. Agrotechnical methods aimed at protecting the slopes are not used in areas with a steep incline, even that exceeding 15%. The only form of their stabilisation is agroforestry, promoted by the authorities, that is, planting trees (e.g. fruit tress) between other plants. Interestingly, according to the 2013 and 2020 Master Plans, terracing of slopes and a version of agroforestry called alley cropping – plant cultivation across the slope, between tree alleys – are also supposed to be supported.

Similar to Havana, an important trait of agriculture in Kigali is intercropping, which is applied in 79% of the analysed agricultural areas. Plants are grown either on adjacent plant beds or (much more frequently) within a single plant bed or a single small field. Maize or yams are dispersed between other plant species (see Figure 5.12). Bananas are planted in a similar manner and additionally provide shade for other, lower plants.

As in Havana, farmers in Kigali primarily use simple agricultural tools, such as hoes, spades or machetes, typical of traditional agriculture with a low level of mechanisation. Artificial fertilisers are also used in the city (in Havana they are

Figure 5.12 Intercropping of yams, sweet potatoes and bananas on a small field on a slope in the *Nyarugenge* district.

Source: Photograph by A. Górna, July 2019.

prohibited); however, access to those agents largely depends on the level of afflu-ence of individual farmers. Some residents cannot afford to buy them. The main importers of fertilisers in the country have their headquarters in the Rwandan capi-tal city; however, high prices frequently limit the possibility for low-income farm-ers to buy those products. To support Rwandans involved in agriculture, national institutions such as the Development Bank of Rwanda and the People's Bank (BPR Bank Rwanda PLC, formerly Banque Populaire du Rwanda SA) offer short-term loans for the purchase of basic production resources. According to an interview conducted with Professor Rufus Jeyakumar[14] from Kigali Independent University, the level of usage of artificial fertilisers in the capital city of Rwanda remains too low to achieve adequate production process efficiency. On the other hand, accord-ing to the provisions of the 2020 Master Plan, excessive use of artificial fertilisers within the administrative borders of Kigali had contributed to soil, surface water and groundwater pollution. For this reason, according to the current document, use of non-organic fertilisers and pesticides is to be limited and replaced by other, eco-friendly practices (City of Kigali 2020). A. Etale and D. C. Drake (2013) point to a different issue, which may prove extremely serious, especially for consum-ers of the food produced in Kigali. According to the authors, some of the valleys within the city are fed by watercourses and canals with water polluted by industrial sewage. An example of a river fed by sewage from textile plants, car workshops and petrol stations located in an industrial zone in *Gikondo* in the north of Kigali is Nyabugogo, located in the north-western part of the research area. Vast agricultural zones stretch along the river, supplying their products to the local market. The results of studies conducted by the authors prove that consumers of those plants (amaranth, sweet potatoes and yams) were exposed to excessive intake of heavy metals, exceeding strict norms adopted for European Union Member States (Etale & Drake 2013). However, it was exclusively a pilot study. It is therefore necessary to plan more comprehensive analyses that would allow for the scale of the problem to be estimated in more detail.

Agricultural areas within the vast valleys of Kigali are divided into smaller fields leased by individual residents (see Figures 5.5 and 5.11). Their number within a single valley may exceed even several hundred, while the area of a single field ranges from several to several hundred square metres. Agricultural zones on the slopes typically take the form of small fields between buildings. An individual parcel is usually cultivated by the family living in its immediate vicinity. In the case of wasteland zones leased from the city, the individuals in charge of a given field may also commute from other parts of the city. The people working in the ana-lysed research areas included both adult men and women; however, estimating the number of farmers proved impossible during the field research without conducting semi-structured interviews. Field observations could not serve as the basis for de-termining the number of employees since the residents work the fields at different times of the day and on different days of the week. Their number also changes continuously depending on the stage of development of the cultivated plants. Nev-ertheless, an important observation made during the field research was the fact that workers in agricultural zones also included children.[15] Involvement of minors in agricultural work is typical of underdeveloped countries. It is estimated that a third of children aged between 5 and 14 take up work in Sub-Saharan Africa, amounting to 68 million working children throughout the continent. In the case of agricultural work, they do not receive any remuneration and their labour is a form of supporting their parents in maintaining the household and securing a living. However, one cannot ignore the fact that using children for hard work in agriculture frequently prevents them from pursuing education, has a negative impact on their physical and mental health and prevents their correct development. According to data from the Rwandan Ministry of Agriculture published in 2016, 8% of all households in Kigali belonged to medium- and high-income farmers, 2% to low-income farmers, 1% to shepherds and 1% to daytime agricultural workers. This means that approximately 12% of households in the city are involved in agriculture (MINAGRI 2016). The remaining groups mainly include vendors and traders (17%), qualified and unqualified workers (24%), craftsmen (13%) and people with a paid job or pursuing their own business activity (31%) (ibidem).

5.4.3 Product distribution and functions fulfilled

Kigali is dominated by subsistence farming. It is characterised by low productivity, low capital expenditure and low level of technical resources and its main purpose is to secure a living for the households. This means that most of the food produced is intended to meet the producers' needs, while the surpluses are sold at local markets. This concerns both small crop fields located in the immediate vicinity of households and vast agricultural zones at the bottom of the valleys. In the case of both groups of agricultural land (small fields and vast valleys), the producer is simultaneously the consumer and the vendor. This is typical of agriculture in Sub-Saharan Africa. The people involved in work at individual fields covered by this research were single residents or entire families, taking care of both production and distribution of food. A clear division of labour is frequently in place between individual family members. Every person is in charge of a specific activity, such as cultivation, harvest, transport of products and their sale.[16] Despite that, Kigali also has a clearly developed network of crop selling intermediaries (Van Dijk & Elings 2014). They purchase fruits and vegetables from the farmers and then supply them to hotels and restaurants or vendors at markets. In many cases, using an intermediary's services is the only way to sell the crops, especially in agricultural areas located far away from the market. Intermediaries also have a considerable impact on the pricing. They can force farmers to sell their fruits and vegetables cheaper, while the farmers, without any other options, have to agree to it. Therefore, intermediaries are another group of actors on the route taken by the products and are frequently the best-earning link in the supply chain, thus contributing to a reduction of profits of the producers themselves (ibidem).

Only five of the analysed agricultural zones feature a sales point located either within their limits or in their immediate vicinity (see Figure 5.13). Crops are typically transported to larger, formal food markets in the city centre (labelled in Figure 5.6) or smaller vegetable stores in the vicinity. The largest marketplaces in

Figure 5.13 A sales point in the immediate vicinity of an agricultural area in the *Gasabo* district.

Source: Photograph by A. Górna, July 2019.

Kigali are the *Kimironko, Kimisagara* and *Nyabugogo* markets. They constitute the main source of fresh agricultural products for the residents of Kigali. Traders typically lease stalls for RWF 20,000 (i.e. approximately PLN 80) per month (ibidem). Every vendor usually distributes the same products every day. Crops from other regions of Rwanda are also distributed at the aforementioned marketplaces, which demonstrates that urban agriculture on its own cannot meet the residents' nutritional needs. Interestingly, the prices of fruits and vegetables in Kigali are set through the vendors' speculations at the *Nyabugogo* market, which take place daily between 3:00 and 6:00 AM. According to a report by N. Van Dijk and A. Elings (2014), the majority of vendors at marketplaces are women.

Apart from the main markets, there are also informal stalls in the city, set up by the residents along major streets in various locations. What is more, some of the crops produced at the analysed fields end up in supermarkets, where they are sold at higher prices than at marketplaces. High-class large stores are gaining prominence in Kigali, primarily as a result of the Rwandan society becoming more affluent. Local agricultural areas also supply the local restaurants and hotels, which typically purchase products directly from the farmers, with omission of any intermediaries, in order to have more control over the quality of purchased fruits and vegetables.

An example depicting the route of the products and the number of actors along the way is one of the agricultural zones analysed during the field research. This small field with an area of 0.13 ha is located on the slope of the *Nyarugenge* hill,

Figure 5.14 A crop field in the *Muhima* sector in the *Nyarugenge* district. On the left, there is a hotel under construction, and on the right – a woman digging up sweet potatoes using a hoe.

Source: Photograph by A. Górna, July 2019.

along KN 1 Avenue in the *Muhima* sector, between the existing ONOMO hotel and a new facility under construction (see Figure 5.14). This is where a semi-structured interview with one of the two women working on the field was conducted. Due to the fact that they spoke neither English nor French, a passer-by who spoke English provided assistance. According to the interview, intercropping of sweet potatoes, potatoes, maize and cassava was practised on the small, fenced-off parcel. Mango trees also grew near the border of the field. During the interview, the women were digging up potatoes, sweet potatoes and cassava using hoes. The parcel is leased temporarily and the crops produced there are intended primarily for sale. It is an example of one of many wasteland zones within the city that are used to produce food with the authorities' approval (excluding industrial-scale production). After the interview ended, two men (most likely intermediaries), following a brief conversation with the women, packed the collected tubers into sacks and transported them on a bicycle to a sales point 700 m away. Although the transport route of the products was short, they went through the hands of intermediaries, who are additional actors on that route. The supply chain, apart from producers and consumers, therefore also involved the manufacturers (as well as, most likely, suppliers) of production resources, intermediaries and finally vendors. Such a high number of actors can increase the final price of the products sold while simultaneously limiting the profits of the producers.

In spatial terms, agriculture in Kigali operates in a shortened supply chain. Fruits and vegetables are not transported over long distances. They typically end up on local marketplaces, situated near the place of production. However, in subjective terms, supply chains of agricultural products are varied. In the case of production within self-supply households, the producer is simultaneously the consumer. The main role fulfilled by urban agriculture in Kigali is that of providing food. Since subsistence farming dominates within the city, its primary aim is to ensure food security of the capital city's population. Nevertheless, in view of the city's development and improved living conditions of its residents, the functions of agriculture are changing as well. Crop cultivation is also an activity that provides additional income to people involved in it. Farmers sell their crops to intermediaries or distribute them themselves at local marketplaces. Thus, urban agriculture is also a form of boosting the modest income of the majority of households. This concerns both farmers themselves and remaining actors in the supply chain – intermediaries and vendors. What is more, taking into account the latest initiatives of Kigali's authorities with regard to promoting kitchen gardens, agriculture in the city is also meant to play an important role in diversifying the residents' diet and popularising healthy nutrition. In Havana and Singapore, urban gardens and farms were also theoretically supposed to fulfil a social and informational function, in addition to food provision. Certain gardens in the Cuban capital city featured cafes that were meant to serve as meeting places for the residents. In Singapore, some of the farms, apart from selling food, also offered tours and trainings, which formed an important part of their activity. Agriculture in Kigali mainly entails other social benefits, such as improvement of the economic situation of the residents, as well as inclusion of excluded individuals (for women, who are engaged in agriculture to a large extent, working on food production and distribution presents a chance for empowerment). However, these benefits tend to be indirect and to some degree are an addition to the basic function of food provision.

5.5 Selected valleys used for agricultural purposes

This sub-chapter contains specifications of four examples of valleys analysed during the field research situated in various parts of the city. Account will be taken of their locations and features of their surroundings, followed by structural and production as well as organisational and technical characteristics. The descriptions are accompanied by profiles of the terrain, depicting the spatial development in each of the valleys (Figures 5.15–5.18). Profile lines were also included in Figure 5.6 (profiles 1–4).

The first of the analysed valleys is situated on the border of two sectors in the *Gasabo* district *Remera* and *Kimironko*, in the north-east part of the research area (see Figures 5.15 and 5.6, profile 1). The length of the designated profile is 800 m. The maximum profile line height is 1,431 m above sea level, while the minimum height is 1,408 m above sea level. The maximum incline is 11.9%, with an average incline value of 5.6%. The highest fragments of the valley slope in the northern part

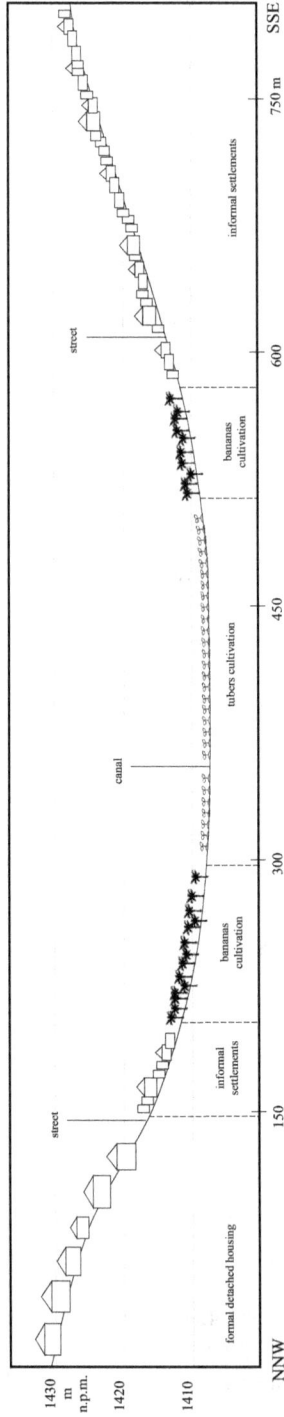

Figure 5.15 Profile of the valley on the border of *Remera* and *Kimornko* sectors in the *Gasabo* district.

Source: Own study.

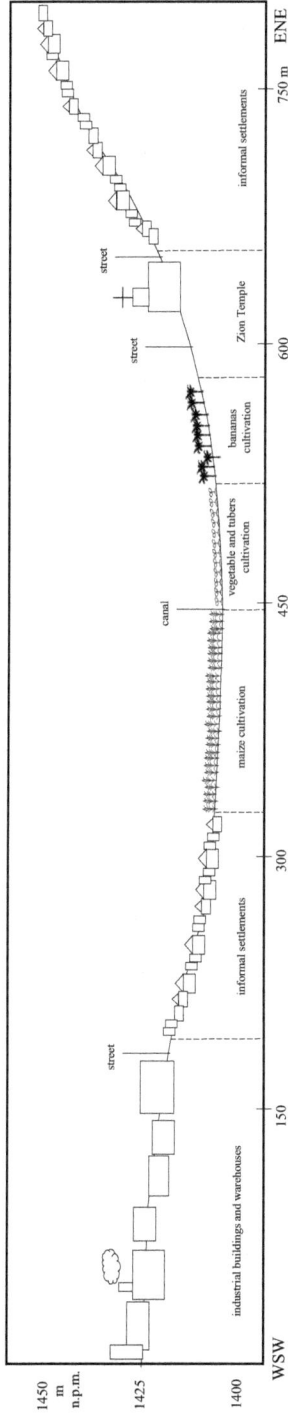

Figure 5.16 Valley profile in the *Gatenga* sector in the *Kicukiro* district.

Source: Own study.

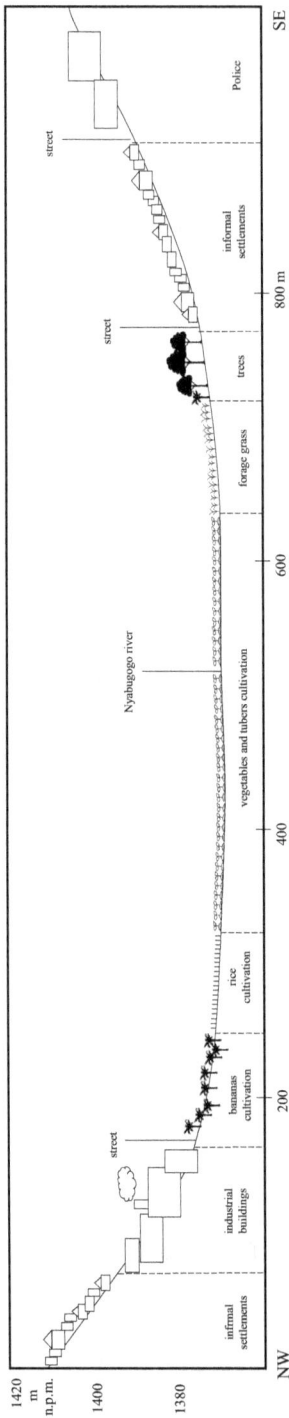

Figure 5.17 Profile of the Nyabugogo river valley on the border of the *Gatsata* sector in the *Gasabo* district and the *Muhima* sector in the *Nyarugenge* district.

Source: Own study.

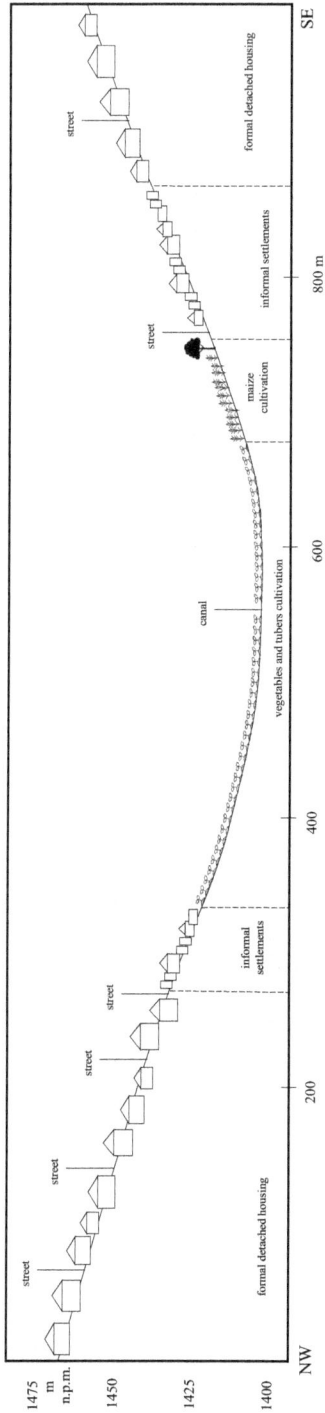

Figure 5.18 Profile of the valley on the border of the *Kimironko* sector in the *Gasabo* district and the *Nyarugunga* sector in the *Kicukiro* district.

Source: Own study.

of the profile are occupied by a neighbourhood of detached houses for the middle and upper class, called *Kibagabaga*, planned on a top-down basis, with a small garden next to every building. Lower fragments of the slope, in the immediate vicinity of arable land, are occupied by informal buildings inhabited by the poorest residents. The building density in this part of the valley is much higher than in the neighbourhood planned on a top-down basis. Informal buildings reach the bottom of the valley, where they border on land used for banana cultivation. The central part of the profile, which is also the bottom of the valley, is primarily dominated by tubers (sweet potatoes, cassava, potatoes) grown on smaller fields, as well as maize, concentrated on the edges. In the southern part of the profile, on the slopes of the valley, bananas are also grown next to the informal buildings located above them.

The second profile line crosses a valley located in the *Gatenga* sector in the *Kicukiro* district (see Figures 5.16 and 5.6, profile 2). The valley's surroundings are dominated by industrial buildings and warehouses, as well as informal residential buildings. To the north of it, there is a vast area on the south-west slopes of the *Kimihurura* hill, which was successively cleared from industrial buildings between April 2019 and September 2020. It currently comprises a number of empty parcels. It was expected that the land would soon be allocated to the construction of a housing estate. However, according to the master plan, the area was deemed dangerous due to the steep incline and natural wetlands are supposed to be recreated there (City of Kigali 2020). The maximum profile line height is 1,449 m above sea level, while the minimum height is 1,409 m above sea level. The average incline is 11.2%, while the maximum incline exceeds 28%. The south-west part of the profile is occupied by industrial buildings and warehouses. It is also, among others, the location of the headquarters of ROKO Construction Limited, a Ugandan construction company, while the majority of warehouses belong to a Rwandan company, MAGERWA Limited. Below, on the slope of the valley, there are informal buildings that reach the bottom of the valley, which is used to grow maize. A small canal runs through the central part of the bottom of the valley; vegetables (e.g. cabbage) and tubers (sweet potatoes and yams) are grown to the south-east of it. Bananas are cultivated farther on, on the slope of the elevation, near the border of the parcel owned by the Zion Temple church. On the other hand, the north-east part of the profile features densely set informal buildings. They occupy terrain with an average incline of more than 17%.

The third profile line crosses the vast Nyabugogo river valley on the border of the *Gatsata* sector in the *Gasabo* district and the *Muhima* sector in the *Nyarugenge* district (see Figures 5.17 and 5.5, profile 3). To the south-west of the valley, there is a vast industrial and commercial zone, which features, among others, one of the largest marketplaces in the city – *Nyabugogo Market*. The maximum profile line height is 1,415 m above sea level, while the minimum height is 1,369 m above sea level. The average profile incline is 6.6%, while the maximum incline exceeds 20%. The steep slopes in the north-west part of the profile are occupied by densely set, informal residential buildings and a small industrial plant located directly next to the street. The area on the other side is occupied by a strip of banana crops. The flat bottom of the valley, in turn, is used to grow rice. Due to high ground moisture content, the rice fields are dispersed throughout the valley. In the central part of the profile, in the vicinity of the Nyabugogo river, vegetables (e.g. cabbage) and tubers (e.g. potatoes, sweet potatoes, yams) are grown on raised plant beds, with individual maize plants dispersed between them. To the south, in an area with the highest moisture, there is a vast field of fodder grass. A row of trees runs along the street situated near the edge of the bottom of the valley. The south-east slopes of the valley are occupied by dense, informal residential and service buildings, with the headquarters of the municipal police station in Kigali in their vicinity.

The last analysed valley is located on the border of the *Kimironko* sector in the *Gasabo* district and the *Nyarugunga* sector in the *Kicukiro* district, in the east part of the research area (see Figures 5.18 and 5.6, profile 4). To the south of the valley, there is the Kigali International Airport. The maximum profile line height is 1,467 m above sea level, while the minimum height is 1,401 m above sea level. The average incline is 13.9%, while the maximum incline slightly exceeds 22%. The south-west part of the profile is the location of one of the housing estates planned on a top-down basis. It is dominated by low-rise detached houses for the middle and upper classes. On the other hand, lower sections of the valley slope are occupied by informal residential buildings. They reach the bottom of the valley and are directly adjacent to arable land. The central part of the profile is dominated by vegetable (e.g. cabbage) and tuber (sweet potatoes, yams) cultivation, while maize and cassava are clustered on the edge of the valley bottom. Similar to the opposite part of the profile, the south-east slopes are occupied by densely packed informal residential buildings (lower fragments) and much sparser formal detached buildings (higher fragments).

All valleys analysed and described above exhibit similar features of the spatial structure, which are also representative of other valleys within the research area. First, the lowest fragments of the slopes are typically taken up by informal residential buildings for the poorest residents of Kigali. The buildings reach the bottom of the valley and are directly adjacent to the crops. These areas are the least suitable for construction due to the steep incline and vicinity of arable land. On the other hand, formal estates of detached houses typically occupy more advantageous parts of the slopes, located higher and with a milder incline. The spatial development of the valleys shows a clear division into zones based on the social status. The placement of crops also exhibits certain commonalities between all analysed agricultural zones. The edges of valley bottom are used primarily to grow maize and bananas, while their central parts, situated on lower, moister ground, are allocated to cultivation of vegetables and tubers. According to the 2020 Master Plan, the bottom of the valleys is supposed to be protected and natural wetlands are meant to be restored to those currently used for crop cultivation. Thus, the informal residential buildings reaching the bottom of the valleys will likely be removed and their residents resettled. The land left after the demolished buildings will most probably be planted over with trees to stabilise the slopes.

5.6   Urban agriculture in Kigali – a summary

Kigali is one of the fastest-developing metropolises on the African continent. Therefore, the city authorities are facing a number of challenges when it comes to ensuring that the continuously increasing population has adequate living conditions, while simultaneously protecting the natural environment, which is particularly at risk of degradation in view of the rate of economic growth and building expansion. The diverse terrain presents an additional impediment for sustainable spatial planning. Nevertheless, it is the presence of several dozen elevations and the vast valleys between them that determines the distribution of urban agriculture in Kigali. It occupies primarily the marshy bottom of the valleys as well as excessively steep hill slopes, which are unsuitable for development. On the one hand, agriculture takes the form of small fields between buildings, while on the other hand – that of vast agricultural zones at the valley bottom, divided into smaller parcels leased by individual residents.

The production structure is dominated by food crops, mainly maize, bananas as well as tubers typical of Sub-Saharan Africa, such as cassava, sweet potatoes or yams. The problem that the city authorities are striving to solve using social campaigns is insufficiently varied diet of the residents. The issue is also reflected in the structure of the crops, with a small share of vegetables, especially leafy greens that are prevalent in Havana and Singapore.

Since Rwanda's capital city is dominated by subsistence farming, the nutritional needs of the residents have a significant impact on its presence within the city space. Merely three decades ago, local food production was the basis for securing the living of the growing Kigali population, which was facing a serious nutritional problem due to the limited supply of crops from the rural areas. At the time, agriculture became a part of the spatial policy of the city authorities and contributed to improving the residents' food security. However, its role is currently changing. First and foremost, since food security has been ensured for a great majority of households in Kigali, agriculture is not so much the basis of livelihood for the residents, but rather an additional source of income. By selling crops, they improve their financial standing. The fact that agriculture is slowly ceasing to be essential to secure the population's livelihood may work to its disadvantage. Once food security of the residents has been achieved, local food production is no longer a strategic activity and is removed to suburban or rural areas, as demonstrated by Singapore's example. However, it should be noted that the food security situation of the residents of the Rwandan capital remains unstable (nearly 35% of households within the city are on the verge of that security). Although urban agriculture is slowly losing its significance, it constitutes somewhat of a safety buffer for the residents.

The spatial and functional structure of Kigali is currently undergoing dynamic changes. On the one hand, they result from the authorities' actions aimed at implementing the principles of sustainable development and the provisions of published planning documents; on the other hand, they are due to increased investments and inflow of foreign capital. As the wealth of the residents of the Rwandan capital city successively increases, urban agriculture will most likely become the first victim of building expansion and growing building density, as demonstrated by the first examples of this phenomenon presented in this chapter. What is more, the authorities' policy towards agricultural zones in Kigali is unclear. On the one hand, they are supposed to be removed from the marshy bottom of the valleys to enable recreation and protection of environmentally valuable wetlands. On the other hand, as part of the policy of improving food security, the authorities are allowing residents to wastelands for crop cultivation for short periods of time.

Since urban agriculture is not a highly profitable activity and Kigali is one of the fastest developing cities in Africa, it is extremely important to ask not *whether* but *how* long agricultural zones will be able to remain within the city's spatial and functional structure. After all, Kigali's authorities are pursuing development directions similar to those currently implemented in Singapore. Therefore, will urban agriculture in Kigali follow in Singapore's footsteps and be gradually removed from the city's central districts to eventually disappear completely from the Rwandan capital?

Notes

1 The analysis was based on an image dated 20 June 2019.
2 Initially, the plan was to visit 100 sites; however, due to the limited time allocated to research works and impeded access to two of the selected crop fields, 98 of them were visited during the field research.
3 This sub-chapter, as well as sub-chapters 5.2–5.6, uses fragments of an original paper: Górna, A., & Górny, K. (2021). Singapore vs. the 'Singapore of Africa'—different approaches to managing urban agriculture. *Land, 10*(9), 987, https://doi.org/10.3390/land10090987 (Creative Commons CC BY 4.0 Licence).
4 The house owned by Richard Kandt is one of the oldest European buildings erected in the territory of modern-day Rwanda. It currently houses the Kandt House Museum of Natural History, where the results of the first research carried out by Kandt and Czekanowski are presented.
5 E. Benken (2017), quoting D. Niyonsenga (2013, p. 2), provides different data, according to which the population of Kigali in 1962 amounted to 55,000, while the area of the town was approximately 3 km².
6 Shortly after gaining formal independence from Belgium, Rwanda quickly fell into the web of neocolonial ambitions of France, whose hegemonic policy in the post-colonial world was focused on all newly emerging Francophone countries. The government in Paris, which supported the Hutu majority's administration, maintained extremely close relations with Rwanda until the 1994 genocide, when the French Foreign Legion humiliated itself during "Opération Turquoise" by hiding the Hutu organisers of the Tutsi massacre. Since Tutsi, led by President Kagame, took over power in the country, the French-Rwandan relations have remained cool while cooperation with the English-speaking world has been developing, as demonstrated by Rwanda's membership in the Commonwealth of Nations since 2009.
7 *Nyamirambo* was established in the 1920s upon the instructions of the Belgian administration. The housing estate was settled mainly by Kiswahili-speaking African domestic servants and traders from the coast of the Indian Ocean. Due to domination of Islam among the residents, *Nyamirambo* is traditionally called the "Muslim district" (de la Croix Tabaro 2014).

8 Until the end of the colonial era, Twa continued to inhabit the forests, leading a hunter-gatherer lifestyle, typical of a society from before the agrarian revolution. They remain a marginalised group in Rwanda's ethnic structure to this day.

9 Niyonsenga (2013; as quoted in Benken 2017) estimates the total population change at 8% (which corresponds to 80%).

10 The interview with the representative of the Kigali City Hall in charge of implementation of the Kigali Master Plan 2013 was conducted in July 2019.

11 Other agricultural zones, not visited in the field, were nevertheless presented in Figure 5.6 to give a more complete illustration of distribution of agriculture within the space of Kigali. Their number in the *Nyarugenge* district amounts to 80, in *Gasabo* to 202, and in *Kicukiro* to 400.

12 In the case of Kigali, it proved impossible to conduct a complex dynamic analysis of distribution of urban agriculture. This is primarily due to the insufficient number of satellite images taken before the year 2019 (i.e. the date of the field research) whose quality would allow for identification of crop fields during a given year. The majority of available images feature noise (i.e. interference associated with the angle of solar rays or the atmosphere, primarily clouds) or have a much lower spatial resolution, preventing manual and visual interpretation within the designated research area, in particular covering all identified agricultural areas.

13 This is a surprising fact, given that practically no animals are bred in the city. Fodder grass is grown in areas unsuitable for cultivation of other plants. One can assume that once harvested, it is transported to suburban areas, where domestic animals are bred.

14 Professor Rufus Jeyakumar, PhD, Dean of the School of Economics and Business Studies, Kigali Independent University, expert on economics and development of agriculture in Rwanda, with particular emphasis of agriculture in cities. The interview was conducted on 27 July 2019.

15 Children working in urban agriculture were also observed during field research conducted in February 2020 in Bissau. They were involved, among others, in fending birds off from rice crops, carrying water and distributing meals.

16 A very similar well-organised labour division system was observed during the field research conducted in February 2020 in Bissau, where urban agriculture was also analysed.

References

Bächtold, P. (2013). *The space-economic transformation of the city*. Heidelberg: Springer, https://doi.org/10.1007/978-94-007-5252-8

Barihuta, P. (2017). Effectiveness of Irondo as a community-led security mechanism in Kigali. *The African Review: A Journal of African Politics, Development and International Affairs*, *44*(1), 62–98.

Barnhart, J. (2011). Umuganda: The ultimate nation-building project? *Pursuit – The Journal of Undergraduate Research at the University of Tennessee*, *2*(1), 3.

Benken, E. E. (2017). *Nowhere to go: Informal settlement eradication in Kigali, Rwanda* (Senior Honors Thesis), https://doi.org/10.18297/honors/127

Bindseil, R. (1988). *Le Rwanda et l'Allemagne depuis le temps de Richard Kandt* [*Rwanda and Germany since the period of Richard Kandt*]. Berlin: DRVB.

Bizimana, J. P., Mugiraneza, T., Twarabamenye, E., & Mukeshimana, M. R. (2012). Land tenure security in informal settlements of Kigali City. Case study in Muhima Sector. *Rwanda Journal*, *25*, 86–100, https://doi.org/10.4314/rj.v25i1.6

Byanafashe, D., & Rutaysire, P. (2016). *History of Rwanda. From the beginning to the end of the twentieth century*. National Unity and Reconciliation Commission.

City of Kigali. (2013). *Kigali City Master Plan Report*, 1–107.

City of Kigali. (2020). *Kigali Master Plan 2050. Master Plan Report*, 1–282.

De Haas, M. (2019). Moving beyond colonial control? Economic forces and shifting migration from Ruanda-Urundi to Buganda, 1920–60. *The Journal of African History*, *60*(3), 379–406, https://doi.org/10.1017/s0021853719001038

de la Croix Tabaro, J. (2014). Know your history: The birth of Nyamirambo, *The New Times*, https://www.newtimes.co.rw/article/109307/know-your-history-the-birth-of-nyamirambo (accessed 4.07.2023).

Etale, A., & Drake, D. C. (2013). Industrial pollution and food safety in Kigali, Rwanda. *International Journal of Environmental Research*, *7*(2), 403–406.

Gazel, H., Harre, D., & Moriconi-Ebrard, F. (2010). *L'urbanisation en Afrique centrale et orientale* [Urbanization in Central and East Africa], Rapport général de l'étude Africapolis II, AFD/e-Geopolis, Paris.

Goodfellow, T. (2014). Rwanda's political settlement and the urban transition: Expropriation, construction and taxation in Kigali. *Journal of Eastern African Studies*, *8*(2), 311–329, https://doi.org/10.1080/17531055.2014.891714

Goodfellow, T. (2017). Urban fortunes and skeleton cityscapes: Real estate and late urbanization in Kigali and Addis Ababa. *International Journal of Urban and Regional Research*, *41*(5), 786–803, https://doi.org/10.1111/1468-2427.12550

Goodfellow, T., & Smith, A. (2013). From urban catastrophe to 'model' city? Politics, security and development in post-conflict Kigali. *Urban Studies*, *50*(15), 3185–3202, https://doi.org/10.1177/0042098013487776

Górna, A., & Górny, K. (2020). Rolnictwo miejskie w miastach Afryki Subsaharyjskiej – ujęcie przestrzenne [Urban agriculture in Sub-Saharan African cities – spatial approach]. *Prace i Studia Geograficzne*, *65*(4), 37–62.

Government of Rwanda. (2005). Organic Law N° 08/2005 of 14/07/2005 Determining the use and management of land in Rwanda.

Government of Rwanda. (2013). Law N° 43/2013 of 16/06/2013 Governing land in Rwanda.

Harvey, D. (2008). The right to the city. *The City Reader*, *6*(1), 23–40.

Hasselskog, M., & Schierenbeck, I. (2015). National policy in local practice: The case of Rwanda. *Third World Quarterly*, *36*(5), 950–966, https://doi.org/10.1080/01436597.2015.1030386

Henninger, S. (2009). Urban climate and air pollution in Kigali, Rwanda. In *The 7th international conference on urban climate*, Vol. 29, 1038–1041.

Henninger, S. M. (2013). When air quality becomes deleterious – a case study for Kigali, Rwanda. *Journal of Environmental Protection*, *4*(8A1), 1–7, https://doi.org/10.4236/jep.2013.48a1001

Ilberg, A. (2008). Beyond paper policies: Planning practice in Kigali. In *9th N-AERUS workshop at the Heriot-Watt University, Edinburgh*, 1–10.

Kinka, H., Onwudiegwu, C., Smeallie, C., & Vajjhala, S. P. (2014). *Kigali, Rwanda: Urban agriculture for food security?* University of Washington, Evans School of Public Policy & Governance, https://doi.org/10.4135/9781526483935

Kuperman, A. J. (2000). Rwanda in retrospect. *Foreign Affairs*, 94–118, https://doi.org/10.2307/20049616

Lefebvre, H. (1968). Right to the city. *Writings on Cities*, 61–181.

Lemarchand, R. (2012). *The dynamics of violence in Central Africa*. University of Pennsylvania Press, https://doi.org/10.1017/s1537592709992556

Manirakiza, V. (2014). Promoting inclusive approaches to address urbanisation challenges in Kigali. *African Review of Economics and Finance*, *6*(1), 161–180.

Manirakiza, V., Mugabe, L., Nsabimana, A., & Nzayirambaho, M. (2019). City profile: Kigali, Rwanda. *Environment and Urbanization ASIA*, *10*(2), 290–307, https://doi.org/10.1177/0975425319867485

Masłoń-Oracz, A., & Mazurewicz, M. (2015). Smart regions and cities supporting cluster development and industrial competitiveness in The European Union. Africa's smart region development influencing global competitiveness. In *Facing the challenges in the European Union*, eds. E. Latoszek, M. Proczek, A. Kłos, M. Pachocka, & E. Osuch-Rak. Warsaw: PECSA, 335–345.

MINAGRI. (2016). *Rwanda 2015. Comprehensive food security and vulnerability analysis*.

Musahara, H., & Huggins, C. (2005). Land reform, land scarcity and post-conflict reconstruction: A case study of Rwanda. In *From the ground up: Land rights, conflict and peace in Sub-Saharan Africa*, eds. C. Huggins & J. Clover. Pretoria: Institute for Security Studies, 270–346.

Nahayo, L., Kalisa, E., & Maniragaba, A. (2019). Awareness on air pollution and risk preparedness among residents in Kigali City of Rwanda. *International Journal of Sustainable Development & World Policy*, *8*(1), 1–9, https://doi.org/10.18488/journal.26.2019.81.1.9

National Institute of Statistics of Rwanda. (2002). *Third population and housing census*, https://www.statistics.gov.rw/datasource/population-and-housing-census (accessed 4.07.2023).

Nduwayezu, G. (2015). Modeling urban growth in Kigali City Rwanda. *Univesity of Twente*.

Nduwayezu, G., Sliuzas, R., & Kuffer, M. (2016). Modeling urban growth in Kigali city Rwanda. *Rwanda Journal*, *1*(1S), 1–31, https://doi.org/10.4314/rj.v1i2s.7d

Nduwayezu, J. B., Ishimwe, T., Niyibizi, A., Ngirabakunzi, B., Nduwayezu, J. B., & Ishimwe, T. N. (2015). Quantification of air pollution in Kigali city and its environmental and socio-economic impact in Rwanda. *American Journal of Environmental Engineering*, *5*(4), 106–119.

Nikuze, A., Sliuzas, R. V., & Flacke, J. (2018). Towards equitable urban residential resettlement in Kigali, Rwanda. In *GIS in sustainable urban planning and management: A global perspective*, eds. M. Van Maarseveen, J. Martinez, & J. Flacke. Taylor & Francis. CRC Press, 325–344, https://doi.org/10.1201/9781315146638-19

Niyonsenga, D. (2013). Urban planning and social inclusion, a study of Kigali city, Rwanda. In *Annual World Bank conference on land and poverty*, Washington, DC.

Nkurunziza, M. (2020). 10 major changes in the new Kigali master plan, *The New Times*, https://www.newtimes.co.rw/news/10-major-changes-new-kigali-master-plan (accessed 18.02.2021).

Oz Architecture and the Conceptual Master Plan Team. (2007). *Kigali conceptual master plan*. Kigali, Rwanda.

Rennie, J. K. (1972). The precolonial kingdom of Rwanda: A reinterpretation. *Transafrican Journal of History*, *2*(2), 11–54.

Rubinoff, D. (2011). *Co-creating Kigali: Working together to create a sustainable city* [Unpublished manuscript]. Kigali.

Rwanda Water Portal. (2021). *Kitchen gardens*, https://waterportal.rwb.rw/toolbox/469 (accessed 21.02.2021).

Seburanga, J. L., Kaplin, B. A., Zhang, Q. X., & Gatesire, T. (2014). Amenity trees and green space structure in urban settlements of Kigali, Rwanda. *Urban Forestry & Urban Greening*, *13*(1), 84–93, https://doi.org/10.1016/j.ufug.2013.08.001

Sirven, P. (1984). *La sous urbanisation et les villes du Rwanda et du Burundi* [Suburbanization and the cities of Rwanda and Burundi], Edition l'Harmattan, Paris.

Sodaro, A. (2011). Politics of the past: Remembering the Rwandan genocide at the Kigali Memorial Centre. In *Curating difficult knowledge*, eds. E. Lehrer, C. E. Milton, & M. E. Patterson. London: Palgrave Macmillan, 72–88, https://doi.org/10.1057/9780230319554_5

Subramanian, R., Kagabo, A. S., Baharane, V., Guhirwa, S., Sindayigaya, C., Malings, C., & Jaramillo, P. (2020). Air pollution in Kigali, Rwanda: Spatial and temporal variability, source contributions, and the impact of car-free Sundays. *Clean Air Journal*, *30*(2), 1–15, https://doi.org/10.17159/caj/2020/30/2.8023

Tadjo, V. (2010). Genocide: The changing landscape of memory in Kigali. *African Identities*, *8*(4), 379–388, https://doi.org/10.1080/14725843.2010.513252

Taguchi, M., & Santini, G. (2019). Urban agriculture in the Global North & South: A perspective from FAO. *Field Actions Science Reports. The Journal of Field Actions*, *20*, 12–17.

The Economist Newspaper. (2012). *Africa's Singapore?*, https://www.economist.com/business/2012/02/25/africas-singapore (accessed 31.01.2019).

Tsinda, A., Abbott, P., Pedley, S., Charles, K., Adogo, J., Okurut, K., & Chenoweth, J. (2013). Challenges to achieving sustainable sanitation in informal settlements of Kigali, Rwanda. *International Journal of Environmental Research and Public Health*, *10*(12), 6939–6954, https://doi.org/10.3390/ijerph10126939

Twahirwa, A. (2018). *Cleanest city in Africa? Kigali scrubs up*, Reuters, https://www.reuters.com/article/us-rwanda-green-pollution-cleanest-city-in-africa-kigali-scrubs-up-idUSKBN1HR2F8 (accessed 31.01.2020).

Uvin, P. (1998). *Aiding violence: The development enterprise in Rwanda*. Kumarian Press, https://doi.org/10.3917/crii.p1999.4n1.0075

Uwayezu, E., & Vries, W. T. D. (2020). Access to affordable houses for the low-income urban dwellers in Kigali: Analysis based on sale prices. *Land*, *9*(3), 1–32, https://doi.org/10.3390/land9030085

Uwimbabazi, P., & Lawrence, R. (2013). Indigenous practice, power and social control: The paradox of the practice of Umuganda in Rwanda in race, power and indigenous knowledge systems. *Interdisciplinary Journal for the Study of the Arts and Humanities in Southern Africa*, *20*(1), 248–272.

Van Dijk, N., & Elings, A. (2014). *Horticulture in Rwanda: Possibilities for further development*. BoP Innovation Center, https://doi.org/10.22059/IJER.2013.619

Wambugu, P. W., & Muthamia, Z. K. (2009). *The state of plant genetic resources for food and agriculture in Kenya*. FAO Commission on Plant Genetic Resources for Food and Agriculture.

Xanthos, D., & Walker, T. R. (2017). International policies to reduce plastic marine pollution from single-use plastics (plastic bags and microbeads): A review. *Marine Pollution Bulletin*, *118*(1-2), 17–26, https://doi.org/10.1016/j.marpolbul.2017.02.048

Záhořík, J. (2011). *Towards demography, labor force and migration in Colonial Ruanda-Urundi*.

6   Summary

6.1   Urban agriculture in Havana, Singapore and Kigali – a comparative analysis

This book focused on analysing the distribution, internal features and functions of urban agriculture in three cities of the Global South: Havana, Singapore and Kigali. This summary will discuss the similarities and differences between the studied metropolises, observed during the conducted research and analyses. It will allow for the specific nature of development of agriculture in various geographical conditions to be determined, as well as for certain universal attributes to be identified.

The selected cities differ both in terms of features of distribution of urban agriculture within their space and in terms of factors affecting the location of agriculture. The first analysed case was Havana, the capital city of Cuba. It is an example of a city where intra-urban agriculture only developed when the residents' food security was seriously threatened. Until the 1990s, the national authorities treated agricultural activity within the capital of Havana as undesirable, as a result of which it was practically non-existent. Only after the collapse of the Soviet Union – the main sponsor of the Cuban regime – in the face of severe food shortage and risk of famine, did urban gardens begin to be established in the city, initially as a grassroots initiative of the residents. They were then quickly included in the national strategy of combating the food crisis. Urban agriculture was institutionalised and incorporated into complex administrative structures. The authorities enabled residents to occupy unused spaces to farm land and establish gardens. Thus, agriculture in Havana emerged within the developed urban tissue by filling the "gaps" in the city's spatial structure. It is currently present in all districts, although its distribution is uneven. The largest number of gardens is concentrated in the parts of Havana characterised by the highest share of free spaces, that is, in the vicinity of the Revolution Square or the military airport, *Aerodromo Ciudad Libertad*. Colonial districts with densely situated historical buildings feature a much smaller number of gardens. Therefore, distribution of agriculture within the space of the Cuban capital can be considered the resultant of the residents' nutritional needs and supply of undeveloped spaces.

In Kigali, in turn, agriculture is concentrated in vast, marshy valleys and on steep hill slopes which have not been developed so far. The terrain plays a key role in the spatial distribution of agricultural activity. Nevertheless, apart from the

DOI: 10.4324/9781003429845-6

landscapes and features of the natural environment, residents in the capital city of Rwanda have a major impact on the distribution of agricultural areas within the city limits, just like in Havana. To improve the food security, especially for the poorest population, the authorities enable Kigali residents to occupy any wasteland zones to produce food that will meet their own needs. It is up to the residents which land within the city is allocated to farming.

The situation in Singapore is different still. The residents' nutritional needs are of little significance when it comes to distribution of urban farms, since the decisions of the authorities are key when it comes to the location of agricultural activity within the city. There are several reasons for this – first, an overwhelming majority of land in Singapore is owned by the state, which gives the authorities practically unlimited control over its use. Due to limited spatial resources and growing building pressure, the policy of the city state in terms of urban agriculture is focused primarily on promoting modern, highly profitable farms, which manage the available space in the most efficient manner possible. Second, food security of the residents is ensured by imports from abroad and is not an issue. Finally, any grassroots initiatives by the residents are significantly limited by the authoritarian government. Urban agriculture in Singapore is concentrated primarily in Kranji Countryside in the north-west part of the country. It is in this poorly connected part of the island, away from the city centre, that the authorities allow for agricultural activity to be located. Apart from that, there are several small farms dispersed throughout the remaining districts.

In Singapore, just like in Kigali, urban agriculture occupies areas that are unsuitable for development. In the former case, this is due to the inconvenient, peripheral location, while in the latter case, it is due to disadvantageous environmental conditions (excessive humidity in valley bottom and excessive slope incline). In Havana, on the other hand, urban gardens are located even in the most central and representative parts of the city, frequently in the vicinity of highly prestigious buildings, for example, in the embassy district of *Miramar*. On the one hand, this demonstrates low building pressure in the city, which has been experiencing economic stagnation for years, similar to the country as a whole; on the other hand, it signifies a serious issue of lack of food security among residents.

Another facet of comparison between the three analysed cities is the structural and production as well as organisational and technical features of agriculture. The plants grown in Havana and Singapore were predominantly vegetables, which is typical of intra-urban agriculture in the majority of cities around the world. Vegetables (especially leafy greens typical of Singapore) spoil easily, so growing them in close proximity to the market is very advantageous. In Kigali, the main plants identified during the field research were maize and bananas, as well as tubers such as cassava, yams or sweet potatoes. Vegetables are grown much less frequently and primarily in valley bottoms. Maize and tubers are the core of the residents' diet in the majority of countries of Sub-Saharan Africa, especially for the poorest and most numerous social groups. This research has shown that in Kigali and Havana, animal husbandry is sporadic. It also accompanied plant crops in all instances. However, in Singapore, animal production was much more prominent. Aquaculture was particularly significant as a result of high share of fish in Singaporean diet and the fact that fish (especially ornamental fish) are important export goods.

The analysed cities are particularly diverse in terms of organisational and technical characteristics. In Havana, the majority of analysed gardens typically utilise organoponics, that is, growing plants on raised plant beds with high compost content. This method is suitable for urban areas with high soil pollution (anthrosols). In Kigali, on the other hand, plants were grown exclusively directly in the ground. In Singapore, traditional crop cultivation directly in the ground is also the dominant method. However, modern hydroponic and aquaponic cultivations promoted by the authorities are also prominent. Those soil-free methods were used by nearly a half of the farms specialising in growing food crops. The important role of high-tech agriculture is primarily the result of access to modern technologies, as well as support provided by the technocratic authorities to farms that use those methods. Contrary to Singapore, Havana and Kigali are dominated by traditional techniques, which do not require high expenditure. In the Asian city state, farms that do not invest in modern technologies and don't have adequately high income are struggling to remain on the market.

Out of the three analysed cases, artificial fertilisers are only used in Kigali. However, access to those fertilisers is limited and largely depends on the farmers' wealth. In Havana, using artificial fertilisers and any types of herbicides and pesticides within the city is completely prohibited. Therefore, only compost and natural methods of protection against pests are used in the Cuban capital. In Singapore, on the other hand, although the issue of using artificial fertilisers is not legally regulated, their use was not confirmed at any of the analysed farms. Several of the visited sites sporadically used herbicides or pesticides, but only those admitted for use by relevant national institutions. Research carried out in Havana and Kigali demonstrated that development of urban agriculture amongst limited resources and low level of farming culture (understood as lack of knowledge, skills and awareness among residents/farmers themselves) entails numerous risks. In the case of the capital city of Cuba, the materials used to build the borders of the so-called *canteros* are asbestos boards and pipes, commonly used in construction of buildings throughout the country. Their use creates a risk of serious health issues, both for food producers and for consumers. In the case of Kigali, a big issue is the crops being located in the bottom of the Nyabugogo river valley, where untreated sewage from nearby industrial plants is released. Food produced within the valley is sold on the local market, used by all residents of the largest city in Rwanda regardless of their economic status. Both in the case of Havana and in the case of Kigali, the lack of adequate measures taken by the authorities to protect the health of producers and consumers is also a serious problem.

The analysed cases are additionally diversified in terms of the transport route of the goods produced by urban farmers. In Havana, agriculture operates in a shortened supply chain, in both spatial and subjective terms. The city is dominated by neighbourhood community gardens run by groups of residents, where the food is sold on the spot or at a nearby marketplace. Therefore, products do not require long-distance transport and are not transported at all in the majority of cases.

Moreover, sale on site allows the producer to directly interact with the consumer. This entails a number of benefits, not only for both actors, but for the urban system itself. First, the consumer gains some amount of control over the production process since they can see "with their own eyes" how the food they buy is produced. Second, this eliminates additional links in the supply chain – intermediaries or carriers, which lowers the final price of the product (consumer benefit), as well as reduces distribution costs (producer benefit). Even the so-called *autoconsumos* create something of small systems or closed supply chains, where most of the food produced is consumed on site by the employees of the work establishment where a given garden operates, with only a small portion being sold or taken by the producers themselves. It should be emphasised that urban gardens in Havana were established primarily at the initiative of the residents and in response to their nutritional needs. Thanks to it, they are located in the vicinity of residential buildings and therefore close to the market. Thus, residents don't have to travel far in order to purchase locally produced food.

The situation in Singapore is different. The most common practice among the analysed sites is sale of products to local supermarkets situated in different parts of the island. However, taking into account the fact that the majority of farms in Singapore are located in Kranji Countryside, a district far away from the city centre, the transport route of the products in extended. Moreover, the number of actors in the supply chain of Singaporean food additionally includes suppliers, intermediaries and vendors. The situation is similar for sites located in the remaining districts, such as *Oh Chin Huat* or *Comcrop*, which distribute their products to supermarkets that can be as far away as about a dozen kilometres away, as well as for farms supplying restaurants in various locations and offering online purchases. In some farms in Singapore, food is sold on site, which allows for direct interactions between the producer and the consumer. However, contrary to Havana, the benefits of the shortened supply chain are significantly limited. Once again, the reason for this is the peripheral location of most Singaporean farms. Residents who wish to purchase food there have to travel long distances in their private cars to the city outskirts, which are unreachable by public transport. Although the distance between the place of production and place of distribution of products is indeed shortened, the distance between the place of distribution and the place of consumption is not. Yet another situation when it comes to urban agriculture can be observed in Kigali. In the majority of analysed agricultural areas, the food produced is used to meet the needs of the producers and their families living in the vicinity (in the case of the kitchen gardens, the place of production is simultaneously the place of consumption). This means that the transport route of the products as well as the distance travelled by the producers themselves are shortened. A frequent practice in Kigali as well as many other cities of Sub-Saharan Africa is sale of surpluses. If the producers themselves are involved in distribution, the value chain is also shortened in subjective terms. However, intermediaries who transport products from agricultural areas to local markets perform an important function in the capital city of Rwanda. Their participation extends the supply chain by an additional link, resulting in an increase in the final price of the product as well as reduced profits of the producers. The three analysed cases demonstrate that although a shortened supply chain is considered a typical feature of urban agriculture in literature on the subject, the situation in many cities is much more complex and depends on a number of local factors.

Varied conclusions can also be drawn with respect to the function of urban agriculture on the basis of the research results presented in this book. In all three instances, urban gardens, farms and crop fields are undoubtedly meant to provide food to residents. However, the situation of residents in the analysed cities is diverse and therefore the role of agriculture itself is also different. In Havana, urban gardens were established in response to a serious threat to food security. Their function has remained unchanged since the 1990s – meeting the nutritional needs of Havana residents. Due to limited supplies of food products from rural areas and from abroad, urban gardens to this day play a key role in the food system of the Cuban capital city. Apart from it, they are also a meeting place for the residents and foster development of interpersonal bonds. However, the social function of gardens is in this case clearly secondary to their nutritional function. Research conducted in Singapore leads to completely different conclusions. Here, the food security of the residents is not at risk. Despite the fact that more than 90% of food products is imported from abroad, the population is not struggling with their limited availability. Singapore is developing dynamically and one of the government's priorities is implementing the principles of sustainable development and building a modern smart city. To improve food self-sufficiency of this global metropolis, the state administration is planning to introduce a number of changes and is beginning to promote local production. Nevertheless, only certain forms of urban agriculture can count on the government's support – primarily highly profitable farms that use modern production methods, which effectively utilise the limited spatial resources, for example, by occupying rooftops of high-rise buildings. Other sites strive to diversify the services they provide in order to stay afloat. Many such farms therefore fulfil additional, for example, recreational and educational, functions. Nevertheless, those which fail to produce calculable economic gains have already been shut down or will be liquidated in the nearest future. In Kigali, similar to Havana, the aim of urban agriculture is primarily to supply the residents with food. After the 1994 genocide, local food production constituted an important element of the strategy of combating the risk of famine. Thanks to it, the residents' situation gradually improved. The majority of households have already achieved food security, while agricultural activity is frequently an additional source of income. Although the situation can hardly be considered fully stable, agriculture is ceasing to be a strategic activity and one should anticipate that its role in the city will be continuously diminishing. The first signs of loss of prominence of urban agriculture were already observed after the end of the field research. Crop fields are currently being successively built over or allocated to other forms of land use, as reflected in the analyses of aerial images. The provisions of Kigali strategic documents, according to which only peri-urban agriculture is supposed to be protected, outline the development paths for the Rwandan capital, in which urban agriculture will be eliminated from the city space. A similar situation is taking place in Singapore, where according to the latest Master Plan, the area occupied by agriculture is supposed to be considerably limited compared with the current state of affairs. The majority of farms in advantageous locations have already been or will shortly be shut down, while Singaporeans who wish to be involved in local food production are looking for solutions that reduce the use of spatial resources. Those include, for example, vertical farms or rooftop farms. The future of urban agriculture is looking the most stable in Havana's case. The city's development is stagnating and additionally slowed down by the recent economic crisis associated with reduced oil deliveries from Venezuela, which has been affected by the presidential crisis since 2019. Unfortunately, this has aggravated the issue of food security of Havana's residents to an extent that has yet to be measured. Therefore, urban gardens in the city continue to play an important role in supplying residents with fruit and vegetables and, as a result, remain within the city space.

The three analysed cases prove – in three different ways – the thesis that urban agriculture, in its traditional form, appears in the city when the residents' food security is at risk and is eliminated when food security is ensured. Havana is an example of a city where the residents are still struggling with the issue of irregular access to food. For this reason, the stable situation of numerous urban gardens is not changing. In Kigali, dynamic development of the city means that the food security of its residents is becoming a less and less urgent problem and the first signs of agriculture being removed from the space of the Rwandan capital city can already be observed. It is being replaced by buildings, as well as other facilities, including recreational ones, such as the golf course in the Gasabo district, described in this book. On the other hand, in Singapore, where food security has been ensured for a long time, traditional agriculture is currently a marginal activity, subject to continuous intense building pressure. However, modern methods such as hydroponics, aquaponics as well as vertical or rooftop agriculture present a chance for development of local food production in this Asian city state. Those methods are becoming increasingly popular and – thanks to the continuously improving efficiency of the production process as well as reduced wastage of resources (e.g. water and power) – they are also promoted by the authorities. Taking into account the fact that for many cities of the Global South, Singapore is a template of a modern, developing city; perhaps similar trends will soon also be observed in other metropolises of the region.

The foregoing conclusions can also be applied in a slightly broader context. On the basis of the research carried out in Havana, Singapore and Kigali, field observations conducted by the author in other cities around the world (Berlin, Kuala Lumpur, Bissau, Dakar, Warsaw, London), as well as the results of works by other researchers, discussed in Chapter 2, urban agriculture can be divided into three types, as proposed in Figure 6.1. Those types were identified based on the functions of urban agriculture and the role it plays within the urban system.

I   The first type is agriculture that serves as the foundation of food supply for residents. It is observed in cities when food security is at risk, typically during economic crises, wars and conflicts or natural disasters, when food supplies to the city are withheld or significantly limited. This type of agriculture normally

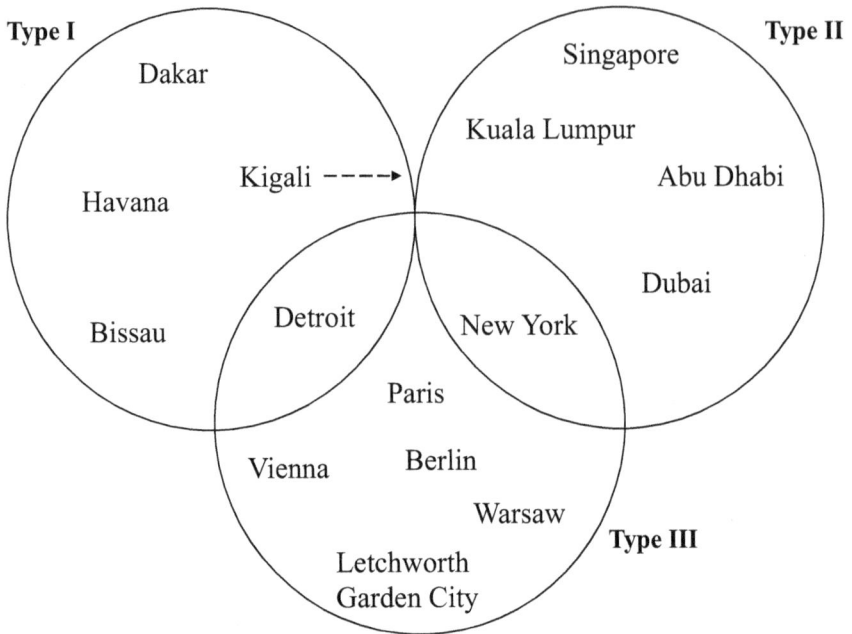

Figure 6.1 Diagram presenting the three types of urban agriculture and cities assigned to those types.

Source: Own study.

utilises simple production methods based on generally available resources. Examples of cities where this type of agriculture has developed include Havana (after the collapse of the USSR and food supplies being withheld to the island), Kigali after the genocide of 1994, as well as Bissau or Dakar, where food security of the residents remains at risk due to the poor economic situation in West African countries.

II The second type is agriculture based on modern technologies and innovative production methods aimed at improving the city's food self-sufficiency. It emerges within the city space not in response to a risk to the residents' food security, but rather as the effect of activity of the most entrepreneurial individuals, research groups and municipal decision-makers seeking optimal solutions to reduce the carbon footprint associated with food production and to use the urban resources (including spatial resources) in the most efficient manner possible. An example of a city that can be assigned to this group is primarily Singapore, as well as other dynamically developing metropolises of the East where solutions aligned with the smart city idea are being implemented, such as Dubai, Abu Dhabi or Kuala Lumpur. Type II agriculture typically does not develop on a wider scale and frequently takes demonstrative or experimental forms. Private companies, as well as foundations and research units play an important role in promoting and implementing this type of agriculture.

III   The third type is agriculture resulting from grassroots initiatives of the residents. Its goal is not the most efficient food production, but rather developing interpersonal bonds, social inclusion of excluded city residents, improving the quality of life in the city, increasing its appeal for tourists as well as education of and raising awareness among residents regarding organic production methods. This type of agriculture is encountered in highly developed cities in democratic countries, for example, European metropolises, such as Vienna, Berlin, Paris, Letchworth Garden City or Warsaw, where community gardens have been gaining popularity and allotment gardens have seen a revival, especially in recent years.

The following diagram presents cities where individual types of urban agriculture have developed.

Apart from the proposed types of agriculture, intermediate types can also be distinguished. Detroit is an example of a city where agriculture combining types I and III has developed. On the one hand, urban gardens in this American metropolis, similar to Havana, were established in response to a serious risk to food security, resulting from the economic crisis which started in 2008. On the other hand, agrihoods[1] established in Detroit, for example, *Michigan Urban Farming Initiative* (founded in 2012), fulfil social, educational and recreational functions. Gardens established in the city are typical community gardens, where the local community produces food to meet its own needs. New York, on the other hand, is a city where intermediate agriculture between types II and III has developed. On the one hand, community gardens are established here as sites of recreation and integration among the residents. On the other hand, the same gardens implement solutions which, based on innovative, pioneering production methods on a global scale, are supposed to increase the food self-sufficiency of New Yorkers.

At the moment, it is difficult to indicate an example of a place where agriculture combining types I and II is developing. However, taking into account the highly dynamic development of Kigali and the fact that municipal decision-makers are implementing solutions within the scope of spatial planning which proved successful in Singapore, perhaps the Rwandan urban agriculture in the upcoming years will undergo transformation and start showing a number of traits characteristic of type II agriculture. The first sign of those changes can be references in the Kigali Master Plan from 2019 regarding use of a modern hydroponic method and rooftop agriculture in food production aimed at meeting the needs of the city residents. The three types of urban agriculture, identified on the basis of the research, can be treated as a post factum thesis. Actual verification of types I, II and III as well as their variants requires, in the first place, a more thorough establishment of the normative framework for each of them, followed by a number of comparative analyses confronting various examples. In view of a growing number of case studies (including those presented in Chapters 3–5 and referenced in Chapter 2) discussing the emergence and functioning of agriculture in various metropolises around the world, as well as its constant variability in places where it has existed for years, it is necessary to seek universal traits of both urban agriculture and the mechanisms determining its development. This is a challenge for further analyses, which appear to be inevitable in view of the many cities turning towards greater sustainability in their interactions with the natural environment.

6.2   Verification of hypotheses and achievement of research objectives

This book formulates three research hypotheses, presented in Chapter 1. They were verified in subsequent Chapters 3, 4 and 5 with respect to the three selected cities (Havana, Singapore and Kigali).

The first hypothesis assumed that the factors affecting distribution of agriculture were the resultant of local socioeconomic, political and environmental conditions. The research carried out in the three cities confirmed the aforementioned assumption in all cases. In Havana, the location of urban gardens was affected, on the one hand, by the nutritional needs of the residents, and on the other hand, by the supply of "gaps" in the urban tissue. It was those "gaps" – unoccupied spaces between buildings – that were transformed into production areas by Havana residents. Districts with a higher supply of such spaces featured more urban gardens. Despite the fact that residents, especially in the 1990s, had freedom to decide on the location of gardens within the city, it should be emphasised that any urban activity within the Cuban capital has been and continues to be subject to strict administrative control. Moreover, apart from the gardens established at the initiative of the people, Havana also features gardens established upon the order of the authorities in the vicinity of various national institutions (e.g. ministries, schools, hospitals or prisons). On the other hand, in Singapore, the location of agriculture is determined primarily by the authorities, which are the mainland owner in the city. They have nearly unlimited control over distribution of agricultural activity on the island. The majority of Singaporean farms are located within the limits of Kranji Countryside, a district designated by the authorities to serve as Singapore's production base. This industrial and agricultural zone is poorly connected with the central parts of the city. In the remaining districts of the city, land is too valuable for it to be allocated to agricultural activity; hence the number of farms is much lower there. Those farms are mainly sites that utilise previously unused spaces in the most efficient manner. This is where, for example, rooftop and vertical farms or mobile gardens are established. In Kigali, there are two main factors affecting the distribution of agriculture. The first one is the terrain. Agricultural activity is mainly concentrated in marshy bottoms of valleys and on steep hill slopes that are unsuitable for development. This factor is also important in other cities of Sub-Saharan Africa, such as Gitega (Burundi), Yaoundé (Cameroon), Brazzaville (Congo) or Kampala (Uganda). In other analysed cities – Havana and Singapore – the terrain proved practically insignificant. The second factor behind the distribution of agriculture in Kigali is the nutritional needs of the residents and freedom regarding the method of satisfying those needs. The population of Rwanda's capital city, similar to the inhabitants of Havana, is allowed to occupy any wasteland zones within the city for crop growing purposes. In each of the case studies, the factors determining the distribution of agriculture within the space of a given city were indicated. Those factors result each time from the local socioeconomic and political conditions, and in the case of Kigali – also environmental conditions. Confirmation of the first hypothesis proves the complexity of the subject matter of research tackled in this book.

According to the second hypothesis, urban agriculture in the analysed cities operates within a shortened supply chain, both in spatial terms (physical distance between the place of production and place of consumption) and in subjective terms (the number of "hands" through which the product passes along its route). This hypothesis was only partly confirmed, which is particularly interesting taking into account the fact that a shortened supply chain is considered a universal and typical trait of urban agriculture in literature on the subject. This was pointed out in Chapter 1, which presents functions fulfilled by agriculture in cities with respect to cases described in literature (on the basis of which this hypothesis was formulated). The results of field research presented in Chapters 3–5 indicate that the hypothesis only proved fully true in Havana's case. In the Cuban capital city, the distance between the place of production and place of distribution and consumption of food is indeed considerably reduced. In the majority of gardens, the products were sold on site to the residents who live in immediate vicinity. The products were not transported and the residents themselves did not need to travel long distances to purchase them. Moreover, in the case of *autoconsumos* gardens, the place of production was simultaneously the place of consumption of the crops. Apart from it, in Havana, in an overwhelming majority of analysed cases, direct interactions between the producer and the consumer were possible and no intermediaries were involved in the supply chain. In addition, production resources such as seeds and compost were typically produced in the gardens themselves or in other locations within the city. Therefore, Havana can be considered a model example of a metropolis where agriculture operates within a shortened supply chain, in many cases reduced to just two links – the producer and the consumer. This entails a number of benefits for the urban system, for example reducing the final price of the product and of distribution costs, as well as increased control over the production process quality. With respect to Singapore, the second hypothesis was partially confirmed. Due to the fact that the city state is forced to import the great majority of food products from abroad, food production within the limits of the island itself considerably shortens the supply chain as it limits the necessity to import goods mainly from Malaysia, Thailand and Indonesia. However, particular note should be taken of local trade in goods produced in Singapore itself and their supply routes should be analysed. In the case of farms selling food on site, the supply chain in subjective terms is reduced to two links, and similar to Havana, the producer interacts directly with the consumer. However, it should be emphasised that the majority of farms that enable purchases on site are located in Kranji Countryside. Due to the peripheral location of those establishments, residents who wish to purchase food on site need to travel long distances. This means that despite the place of production being simultaneously the place of distribution, the distance between the place of distribution and the place of consumption is considerable, which limits the benefits derived from a shortened supply chain, in particular those associated with reducing the carbon footprint during transport of products. With respect to Kigali, the second hypothesis was also only partially confirmed. In the majority of analysed agricultural areas, the food produced is used to meet the needs of the producers themselves. In this case, the supply chain should be considered completely reduced in both subjective terms (the producer is simultaneously the consumer) and in spatial terms (producers typically live in the neighbourhood or in close proximity to the fields they farm). However, selling production surpluses in local marketplaces is a common practice in the Rwandan capital city. Since the distance between the agricultural areas and the marketplaces is short, the products are transported to sales points on foot or by bicycle. In spatial terms, the supply chain is also reduced in this instance since the crops are ultimately sold from fields in one district to houses in neighbouring parts of the city. However, the supply chain is in many cases extended in subjective terms. Intermediaries involved in transport of products from agricultural areas and their supply to vendors at local marketplaces play an important role in Kigali. They significantly increase the final product price and reduce the profits of the producers. Actors along the supply chain in Kigali include suppliers of production resources (seeds, fertilisers), producers, intermediaries, vendors and, finally, consumers. Each subsequent link lowers the profits of the producers, who are frequently in a difficult financial position since the people involved in agriculture in Kigali belong to the poorest social class.

The third hypothesis assumed that urban agriculture fulfils different functions, depending on the socioeconomic and political conditions in a given city/country. The field research described above allowed for full confirmation of this hypothesis. In the three analysed cities, which are characterised by different degrees of socioeconomic development, living standards of their residents as well as policies pursued by the authorities, the roles of urban agriculture are different. In Havana, where access to food products has been limited since the 1990s, the priority of urban gardens is improving the residents' food security. In addition, according to the authorities' policy, agriculture is also supposed to fulfil social, educational and recreational functions (the *La Ceiba* garden). In Singapore, although food security of the residents is ensured, the need to import food from abroad is a significant problem. The role of urban agriculture in this city state is not increasing the supply of food to meet the residents' basic needs, as it is the case in Havana and Kigali, but primarily to increase the share of local production, which would improve the city's food self-sufficiency. What is more, the majority of farms in Singapore should be considered multifunctional enterprises, primarily due to the fact that many of them find it difficult to remain on the market by selling food alone. Expanding the offering by adding more services is therefore a part of the strategy for improving their financial standing. Apart from food production, farms also fulfil educational, tourist or recreational functions. Contrary to Havana, in Singapore's case, the multifunctionality is an attempt at remaining on the market, not a top-down initiative of the authorities or an indirect result of the way gardens in Havana operate (points of sale are simultaneously meeting places for the residents, supporting development of interpersonal bonds). In Kigali, similar to Havana, the aim of urban agriculture is first and foremost ensuring the food security of the residents. Apart from this, it serves as an extra source of income for some Kigali residents while simultaneously playing an important role in the strategy of securing their livelihood. Although the current nutritional needs in the majority of households in the city are met, the situation remains unstable. Urban agriculture still fulfils the basic role of food provision and its functions are not diversified. Taking into account the fact that the Rwandan capital is one of the most dynamically developing cities on the continent, the role of agriculture in the urban system will be changing in the nearest future due to investments in the development of infrastructure, expansion and increasing building density, as well as the growing wealth of the residents. Urban agriculture in Kigali will start gradually becoming a marginal activity and will cease to play such an important role in the urban system, similar to what is currently the case in Singapore.

Thanks to research carried out for the purposes of this book, the five research objectives were achieved. In all three cities, the location of urban agriculture, features of its distribution and factors affecting it were identified. Structural and production as well as organisational and technical features of individual urban gardens and farms were also specified along with the functions they fulfil. At some of the analysed sites, obtaining answers to all research questions was impeded for various reasons. This was primarily due to the lack of possibility to conduct interviews in some locations (e.g. gardens within the limits of the State Security Department *Villa Marista G-2* in Havana), as well as unwillingness on the part of some respondents to share the details of a certain garden or farm's operations. At certain sites in Singapore, owners, despite being assured of academic purposes of the research, were afraid of disclosure of strategic information to their competitors. In Havana, on the other hand, certain respondents consistently avoided answering questions about control by national authorities.

At this point, it should be mentioned that conducting field research in the three selected cities had its limitations, which affected the final scope of obtained information and data. In the case of Havana, the most serious difficulties were associated with the state administration's unwillingness to grant permits that would allow for research to be freely conducted in urban gardens. For this reason, the success of the field research largely depended on the politeness of the residents of Havana themselves. Fortunately, the majority of representatives of the gardens were willing to participate in the proposed interviews, which were not recorded for practical reasons. However, there were instances where employees did not permit entry to the garden area and refused to answer any questions, justifying their decision with the authorities' instructions. In Singapore, reaching some of the respondents proved to be a major difficulty due to the fact that meetings with some farm representatives could not be arranged before travelling to conduct the field research. In those instances, it was necessary to look for respondents next to entryways to private properties, which proved impossible several times. Despite the fact that conducting research in Singapore should be considered relatively safe (the crime rate in the city is one of the lowest in the world), in Kranji Countryside, groups of freely running stray dogs proved a major impediment. The research had to be interrupted twice due to immediate risk of being bitten. In Kigali, on the other hand, because of the varied terrain, it proved necessary to conduct research using a leased off-road vehicle, which had not been anticipated earlier. The problem was the condition of the roads, which frequently made it difficult to reach previously designated agricultural areas. What is more, in several locations, it was impossible to collect photographic documentation. Residents did not always agree to their crop fields being photographed. Moreover, due to the language barrier in Kigali, it proved difficult to conduct semi-structured interviews since the more impoverished people engaged in agriculture speak Kinyarwanda. Despite the aforementioned difficulties, attempts were made at achieving the highest possible efficiency of the conducted field research by adjusting the research methods and techniques each time to the local conditions.

6.3   Urban agriculture – future research perspectives

Urban agriculture is a prominent subject in research on the development of contemporary cities of the Global South, as demonstrated by the multitude of studies in international academic literature. However, it should be emphasised that this issue has thus far been treated selectively. Many metropolises have not been analysed in terms of distribution of agriculture, its internal features as well as the functions it fulfils in the spatial and functional structure. On the one hand, it is therefore necessary to also conduct comprehensive research on other cities of the Global South, especially less frequently analysed hubs which are not capital cities. The processes taking place there could differ from those analysed thus far in the capital cities, which are typically subject to particular building pressure and faster modernisation processes. On the other hand, apart from improving knowledge on urban agriculture by adding new case studies, there is also a clear need for comparative and synthesising work, which would allow for more general conclusions to be drawn and for the phenomena taking place at a micro scale to be put in a broader context.

The research carried out in Havana also brought to light a major problem associated with the development of urban agriculture amongst limited resources. The ubiquity of asbestos boards in gardens using the organoponic method in the Cuban capital city is highly concerning. Therefore, comprehensive research within the scope of agrotechnology is necessary to determine the harmful substance content in the soil and plants, as well as within the scope of medicine to determine the effects of this common yet dangerous practice on the health of producers and consumers.

Moreover, taking into account the fact that urban agriculture can present a chance for sustainable development not only for the cities of the Global South but also hubs in highly developed countries, it is also necessary to carry out research to indicate good and bad practices in various geographic contexts. They should concern issues such as: methods of mitigating pollution absorption by food crops, optimal use of municipal waste in compost production and treated sewage in crop irrigation. What is more, there appears to be a need for studies concerning development of sustainable food systems based on a shortened supply chain as well as comprehensive analyses of the efficiency of modern production methods, such as those applied in Singapore. Urban decision-makers are currently searching for effective methods of mitigating the negative impact of climate change, air pollution, excessive spatial expansion of urban buildings as well as measures that would contribute to improving food security of the residents. Research presenting solutions which proved effective and beneficial for the population as well as the urban ecosystem in a specific socioeconomic, political and environmental context could be implemented in other hubs, contributing to the development of sustainable, self-sufficient cities.

Note

1 Agrihood – a neighbourhood integrating the residential function and urban agriculture. The term agrihood is a blend of two words: agriculture and neighbourhood.

Index

Note: *Italic* page numbers refer to figures and page numbers followed by "n" denote endnotes.

For Product Safety Concerns and Information please contact our EU
representative GPSR@taylorandfrancis.com
Taylor & Francis Verlag GmbH, Kaufingerstraße 24, 80331 München, Germany